国家出版基金项目

中国古建筑测绘大系·宗教建筑

洪洞建筑群

清华大学建筑学院 编写

王贵祥 刘畅 贾珺 廖慧农 王南 主编

"十二五"国家重点图书出版规划项目

中国建筑工业出版社

Traditional Chinese Architecture Surveying and
Mapping Series:
Religious Architecture

ARCHITECTURE COMPLEX OF

HONGTONG

Compiled by School of Architecture, Tsinghua University
Edited by WANG Guixiang, LIU Chang, JIA Jun,
LIAO Huinong, WANG Nan

China Architecture & Building Press

Contents

Introduction 001

Preface 005

Figures 015

Upper Guangsheng Monastery 017

Water God's Temple 119

Lower Guangsheng Monastery 155

Name List of Participants Involved in Surveying and
Related Works 227

目　录

参与测绘及相关工作的人员名单——二二六

广胜下寺——一五五

水神庙——一一九

广胜上寺——〇一七

图版——〇一五

序言——〇〇五

导言——〇〇一

Introduction

Since its inception in 1946, the School of Architecture at Tsinghua University has been committed to surveying and mapping traditional Chinese buildings, following the practice of the Society for the Study of Chinese Architecture (*Zhongguo Yingzao Xueshe*) that LIANG Sicheng, a driving force in the Society and founder of Tsinghua's architecture department (known as the School of Architecture since 1988), and his assistant MO Zongjiang brought with them to Tsinghua. Between 1930 and 1945, with members of the Society, LIANG visited over two thousand Chinese sites located in more than two hundred counties and fifteen provinces, and discovered, identified and mapped over two hundred groups of traditional buildings, including the famous Tang-period east hall (dating to 857) of Foguang Monastery at Mount Wutai—which was not an easy task because of the harsh working conditions in the secluded and relatively inaccessible villages in the countryside. In that same spirit, despite the difficult political circumstances from the 1950s through the 1970s, the School of Architecture conducted a systematic survey of historical buildings in the New Summer Palace (Yiheyuan). At the beginning of the Cultural Revolution in the late 1970s, all members of the faculty focusing on the history of architecture went to Hebei province under the leadership of MO Zongjiang to measure and draw the main hall of Geyuan Monastery in Laiyuan, an important Liao-period relic hidden in the remote mountains. This was followed by in-depth research and analysis. At the same time, those professors that specialized in Chinese architectural history (MO Zongjiang, XU Bo'an, LOU Qingxi, ZHANG Jingxian, and GUO Daiheng) led a group of graduate students to Zhengding in Hebei province, where they conducted component analysis and research of Moni Hall at Longxing Monastery, a Northern-Song timber-frame structure that had partially collapsed but was then in the process of being rebuilt. They also investigated nearby

导　言

因为前辈学者梁思成及其助手莫宗江两位先生从中国营造学社继承的传统，清华大学建筑学院自创立以来，一直十分注重古代建筑实例的实地考察与测绘。尽管在 20 世纪 50 至 70 年代受到各种因素的影响与冲击，那时的清华大学建筑系，还坚持了对颐和园内一批古代建筑实例的系统测绘。改革开放刚刚开始的 1970 年代末，清华大学建筑历史方向的全体教师，就在莫宗江先生带领下，共同远赴偏僻的河北山区，考察测绘了创建于辽代的涞源阁院寺大殿，并对这座辽代木构建筑进行了系统研究。同是在那一时期，建筑历史教研室的莫宗江、徐伯安、楼庆西、张静娴、郭黛姮等教师，带领研究生赴河北正定，除了对正在落架重修的北宋木构大殿隆兴寺摩尼殿的大木构件进行现场分析研究外，还对正定及周边的古建筑进行了系统考察与调研。这种由老先生带队，

historical buildings in and around Zhengding. This practice of teamwork—senior researchers, instructors, and (graduate) students participating in the investigation and mapping of traditional Chinese architecture side by side—became an academic tradition at the School of Architecture of Tsinghua University.

Since the 1980s, fieldwork has been a crucial part of undergraduate education at the School, and focus and quality of teaching has constantly improved over the past decades. In the 1990s, professors like CHEN Zhihua and LOU Qingxi carried out surveying and mapping in advance of (re)construction or land development on sites all across China that were endangered. Since the turn of the twenty-first century, the two-fold approach—attaching equal importance to practice (fieldwork) and theory (teaching)—was widened and deepened. Sites were deliberately chosen to maximize educational outcome, resulting in a broader geographical scope and spectrum of building types. In addition to expanding on the idea of vernacular architecture, special attention was paid to local (government-sponsored) construction of palaces, tombs, and temples built in the official style (guangshi) or on a large scale (dashi), and to modern architecture dating to the period between 1840 and 1949. Students and staff have accumulated a lot of experience and created high-quality drawings through this fieldwork.

In retrospect, we have completed surveys of several hundred monuments and sites built in the official dynastic styles of the Song(Jin), Yuan, Ming and Qing all across the country. Fieldwork was always combined with teaching. Among the architecture surveyed are the (single- and multi-story) buildings in front and on the sides of the Hall of Supreme Harmony in the Forbidden City in Beijing; the architecture at Changling, the mausoleum of emperor Jiaqing located at the Western Qing tombs in Yi county, Hebei province; the monasteries on Mount Wutai, Shanxi province, including Xiantongsi, Tayuansi, Luobingsi, Pusading, Nanshansi (Youguosi), and Longquansi; Zhongyue Temple, Songyang Academy, and Shaolin Monastery in Dengfeng, Henan province; Xiyue Temple, Yuquan Court, and the Taoist architecture on the peaks of Mount Hua in Weinan, Shaanxi province; Chongan Monastery, Nanjixiang Monastery, Jade Emperor Temple (Yuhuangmiao) in Shizhang, and the temples of the Two Transcendents (Erxianmiao) in Xiaohuiling and Nanshentou, all situated in Lingchuan county of Shanxi province; and the upper and lower Guangsheng monasteries and the Water God's Temple in Hongdong, Shanxi province. In recent years, we have developed a specialized interest in the study of religious architecture of Shanxi province and investigated almost a dozen privately- or government-sponsored Song and Jin sites

002

教师与研究生集体参与，对古代建筑进行深入考察与测绘研究的做法，在清华大学形成了一个良好的学术传统。

1980年代以来，清华大学建筑学院始终在本科教学环节中，坚持讲授古代建筑测绘这门经典课程。这一传统在21世纪初的这十几年中始终延续。如果说，20世纪90年代由陈志华、楼庆西等教授带领的测绘教学，将相当的注意力放在了分布于全国多个省、市、自治区大量传统乡土村落建筑的抢救性测绘上，进入21世纪以来，清华大学建筑学院开展的这种结合本科教学的古建筑测绘教学与实践，覆盖的地域范围与建筑类型范围更为宽广：除了进一步拓展乡土建筑的测绘之外，在对各地留存的历代官式或大式建筑，如宫殿、陵寝、寺庙等建筑的测绘以及近代建筑的测绘上，也积累了大量测绘经验、图纸及丰富的调研资料。以古代官式建筑测绘为例，结合本科教学，我们先后完成了北京故宫太和殿前及两侧门殿、楼阁与朝房建筑，河北易县清西陵昌陵完整建筑群，山西五台山显通寺、塔院寺、罗睺寺、菩萨顶、南山寺（佑国寺）、龙泉寺等多座组寺院建筑群，河南登封中岳庙、嵩阳书院、少林寺古建筑群，陕西渭南华山西岳庙、玉泉院及华山山顶各道观古建筑群，山西陵川崇安寺、南吉祥寺、小会岭二仙庙、南神头二仙庙、石掌玉皇庙，以及山西洪洞广胜上寺、广胜下寺、水神庙等数百座古建筑实例的测绘，其时代的范围覆盖了宋（金）、元、明、清等历代木构建筑遗存实例。近几年，我们又将测绘的重点放在了高平、晋城等晋中及晋东南地区，

located in central Shanxi (Jinzhong) and southeastern Shanxi (Jindongnan), specifically in Gaoping and Jincheng counties. This includes the Youxian, Chongming, and Kaihua monasteries and the Two Transcendents Temple in Xilimen. Additionally, supported by the State Administration of Cultural Relics, the head of the Architecture History and Historic Preservation Research Institute at the School of Architecture, Liu Chang, led a group of students to map and draw the main hall of Zhenguo Monastery in Pingyao, a rare example from the Five Dynasties period. The survey results have been published. Tsinghua fieldwork in Shanxi has become an annual event that is jointly organized almost every summer by the faculty of the School of Architecture, including professors engaged in research on non-Chinese architecture, in cooperation with their graduate students.

It is worth mentioning that since 2007, the School has worked in collaboration with the well-known company China Resources Snow Breweries Ltd., which supports the transmission and dissemination of knowledge on traditional Chinese architecture and provides funds for the School's research and field investigation activities. Drawing on the support from industry allowed us greater initiative and flexibility, and we were thus able to carry out research on and survey often overlooked but no less important Song-Jin monuments in central and southeastern Shanxi.

Our years-long fieldwork has not only enabled us to teach students subject knowledge about scale, material, form, and decoration of traditional Chinese architecture as well as a sense of appreciation for the old, but has also provided us with plenty of data for monument preservation practice and research. China Architecture and Building Press spared no effort in compiling and publishing the results of the fieldwork in 2012. Publication has also been supported by the National Publishing Fund. This highlights not only the importance of our contribution to architectural education at the national level but also shows its significance for the transmission, development, and revival of traditional Chinese architectural culture both at home and abroad. In order to expand the reach of this work to an international audience, *the Traditional Chinese Architecture Surveying and Mapping Series* is being published bilingually. Based on the past ten years of fieldwork, we have now compiled five volumes, namely *Mount Wutai's Buddhist Architecture* (Traditional architecture on Mount Wutai, Shanxi), *Architecture Complex of Songshan* (Traditional architecture in Dengfeng, Henan), *Mount Hua's Yuemiao and Taoist Temples* (Traditional architecture on Mount Hua, Shaanxi), *Architecture Complex of Hongtong* (Traditional architecture in Hongtong, Shanxi), and *Architecture Complex*

对包括高平游仙寺、崇明寺、开化寺、西李门二仙庙等在内的十余座宋金建筑群，进行了全面而系统的测绘。这一期间，在国家文物局的支持下，建筑历史与文物保护研究所刘畅老师还带领研究生对五代时期创建的平遥镇国寺大殿等建筑进行了精细测绘，并出版了测绘研究成果。此外，清华大学建筑学院的测绘工作，几乎每年都是由全体建筑历史教师共同合作，并带领研究生们共同完成的。从事外国建筑史教学的老师，也不例外。

特别值得一提的是，自 2007 年以来，清华大学建筑学院与国家知名企业华润雪花啤酒（中国）有限公司建立了良好的合作关系。该集团不仅支持中国古建筑知识的传承与普及工作，也对清华大学建筑学院中国古代建筑研究及古建筑测绘工作给予了直接的支持，使得我们的古建筑测绘工作变得更为主动和更具选择性。一大批珍贵的山西晋中及晋东南地区宋金时代建筑实例的测绘与研究，就是在这样一个前提下得以顺利开展与完成的。

坚持数十年的古建筑测绘工作，不仅在培养学生对传统中国建筑的尺度、材料、造型与细部装饰的认知与感觉上起到了直接的影响，而且也为各地文物建筑保护与研究工作，提供了相当充分的资料支持。

2012 年，中国建筑工业出版社花大气力组织了汇集全国重点院校建筑系古建筑测绘成果的中国古代建筑测绘大系的编辑出版工作。这一工作也获得了国家出版基金的支持。这不仅是对高校建筑教育成果的一份支持，也是对中国传统建筑文化传承、发展与复兴的一份支持。正是在这样一个背景与前提下，我们对近十余年来考察测绘的古代建筑案例加以整理，分别编汇了包括《五台山佛教建筑》《嵩山建筑群》《华山岳庙与道观》《洪洞建筑群》《高平建筑群》5 册古建筑测

of Gaoping (Traditional architecture in Gaoping, Shanxi). The architectural drawings presented in these books are carefully selected and screened by Tsinghua professors. They only show a part of our comprehensive surveying and mapping work, but still cover a whole spectrum of geographic regions and time periods. Thus, they contain information of high academic value that may serve as a reference for future study and for the protection of cultural heritage. It is hoped that our work will help to promote interest in and improve understanding of traditional Chinese architecture, not only among Tsinghua students (through hands-on experiences in the fieldwork) but also among architectural historians and professionals engaged in monument preservation at home and abroad.

As a final thought, let me shortly address the workflow. The drawings presented here are based on survey and working sketches drawn up on site during several years of fieldwork conducted by Tsinghua professors together with graduate and undergraduate students. Back home, the measured drawings were redrawn over months of diligent work by graduate students with computer-aided software to achieve dimensionally accurate and visually appealing results, a project that was completed under the supervision of LIU Chang, head of Tsinghua's Architecture History Institute, and the Tsinghua professors LIAO Huinong and WANG Nan, as well as TANG Henglu and his colleagues from the WANG Guixiang Studio. We would like to take this opportunity to thank the professors, students and colleagues who participated in the fieldwork and its revision.

Our final thanks go to LI Jing, assistant researcher at the Architecture History Institute here at Tsinghua. Next to participating in surveying and mapping, she organized the development of the book and moreover, made this book possible in the first place.

WANG Guixiang, LIU Chang, LIAO Huinong
Architecture History and Historic Preservation Research Institute, School of Architecture,
Tsinghua University
December 5, 2017

Translated by Alexandra Harrer

绘图集，作为这套『中国古建筑测绘大系』的部分成果。尽管这只是我们多年测绘成果的一部分，但也是清华建筑历史学科教师们仔细筛选、认真校对、充分整理之后的较具典型性与参考性的成果。

这些成果不仅地域覆盖面大，而且建筑遗存的时代跨度也相当长，具有十分重要的学术价值。希望这些成果对高校建筑系学生们学习古建筑，建筑历史学者研究古建筑，以及文物保护工作者从事文物古建筑的保护与修缮，能够起到积极的推动作用与重要的参考价值。

最后要提到的一点是，除了参与测绘的教师、研究生与本科生多年历尽辛苦的测量与绘图工作之外，此次清华大学建筑学院承担的这5册测绘图集，也经由建筑历史与文物保护研究所刘畅、廖慧农、王南和他们的研究生，以及王贵祥工作室团队的唐恒鲁等同仁们在既有测绘图纸基础上，经过数月认真仔细的线条分层、图面调整、数据校对、图面完善等缜密修复工作，在这里也要向参加测绘图整理的老师、同学和同事们表示感谢。

还应该特别提到的是建筑学院建筑历史与文物保护研究所的助理研究员李菁博士，她不仅参加了多次测绘，还为这套书最后的编辑与出版做了大量相关工作。这里一并表示感谢。

清华大学建筑学院 建筑历史与文物保护研究所

王贵祥、刘畅、廖慧农

2017 年 12 月 5 日

Preface

The architecture complex of Hongtong, located at the southern foot of Huo Mountain seventeen kilometers northeast of the county-city of Hongtong (formerly Zhaocheng), Linfen prefecture, Shanxi province, is an important monument of traditional Chinese architecture from the Yuan and Ming dynasties, and formed. The architecture complex of Hongtong consists of three parts: the upper monastery situated atop a hill at the southwestern corner of Taiyue Summit, the main peak of Huo Mountain; the lower monastery on the foothill; and the Water God's Temple (Shenshuimiao) dedicated to the spirit of Huo Spring known as Mingying Wang. The timber-frame buildings from the Yuan dynasty, the towering glazed-tile Feihong (Flying Rainbow) Pagoda from the Ming dynasty, and the numerous statues and murals inside the halls are all rare cultural treasures (Fig.1). LIN Whei-Yin and LIANG Ssu-ch'eng, Chinese architectural historians of the First Generation, once poignantly said:

"The murals in Mingying Wang Hall and the beam frameworks of the upper and lower monasteries' halls are extremely precious and rare relics; they are outstanding examples of what we have not seen elsewhere before. From the viewpoint of art history, they provide crucial historical evidence." [1]

Guangsheng Monastery is situated in a quiet and beautiful spot in a hilly landscape with clear spring water and picturesque cypress tress. According to legend, the complex was built in 147 (the first year of the Jianhe reign period of the Eastern Han dynasty). Then known as King Asoka Pagoda Court and Julushe Monastery (named after a Buddhist unit of distance), it was designed to house a relic of king Asoka stored in a pagoda. In the Baoding reign (561-565) of the Northern Zhou dynasty, the eminent monk Zheng Jue mentioned the by-then already dilapidated condition of the pagoda

[1] LIN Whei-Yin, LIANG Ssu-ch'eng. A survey record of the traditional architecture in Fen region, Shanxi province (Jin Fen gujianzhu yucha jilüe)[J]. Journal of the Society for Research in Chinese Architecture (Zhongguo yingzao xueshe huikan), 1935, 5(3).

序　言

洪洞建筑群位于山西省临汾市洪洞县（原属赵城县）东北17公里的霍山南麓，为中国重要的元、明建筑群遗物。洪洞建筑群包括三部分：位于山顶（太岳主峰西南隅小山之巅）的广胜上寺、位于山麓的广胜下寺和供奉霍泉水神的水神庙。三座建筑群中的多处元代木构，高耸云霄的明代琉璃塔——飞虹塔，以及诸殿宇内的塑像与壁画，皆为难得之瑰宝（图一）。林徽因、梁思成在二人合著的《晋汾古建筑预查纪略》一文中曾经盛赞洪洞建筑群曰：『明应王殿的壁画，和上下寺的梁架，都是极罕贵的遗物，都是我们所未见过的独例。由美术史上看来，都是绝端重要的史料。』[1]

[1] 林徽因，梁思成．晋汾古建筑预查纪略[J]．中国营造学社汇刊，1935，5（3）．

图一 《平阳府志》中的广胜寺（来源：《洪洞广胜寺》，2006年）

Fig.1 Guangsheng Monastery in *Pingyang Fuzhi*. Source: *Hongtong Guangshengsi*, 2006

and expressed his hope for future repair. During the Dali reign (766-779) of the Tang dynasty, general GUO Ziyi presented a memorial to the emperor in which he suggested reconstruction after having seen the broken buildings of the monastery—the ancient pagoda was almost razed to the ground. We can not exactly know what really happened in the subsequent years, but Song poetry (eyewitness testimony) suggests an impressive (new) layout of the monastery in that period. The monastery was slightly expanded in the Jin dynasty but retained the original Tang-Song layout. The murals at the Water God's Temple provide some basic information about the historical design back then— the upper monastery followed a pattern with a pagoda (*ta*) in the front and a multi-story pavilion (*ge*) in the back with gates (front gate [*shanmen*] and gate to the pagoda court [*tayuanmen*]) aligned on the central axis one after the other (Fig.2). The monastery's current condition pales in comparison to the grandure of its former state during this period. The monastery was ruined in the Mongol-Jin wars during the Zhenyou reign (1213-1217) of the Jin emperor Xuanzong. In the early Yuan period, Guangsheng Monastery and the Water God's Temple were rebuilt in a magnificent palatial style with ritual halls and utility buildings measuring up to several hundred bay-lengths. Unfortunately, the entire monastery was destroyed again in an earthquake in 1303 (the seventh year of the Dade reign period of the Yuan dynasty), but re-erected shortly thereafter. Most wooden buildings date to this reconstruction. The Ming dynasty, in addition to regular maintenance and repair of the halls, rebuilt Feihong Pagoda.[2]

1. Upper Guangsheng Monastery

Situated atop a hill, the architecture is fitted into the existing landscape, resulting in a slightly curved main axis along which the important buildings are aligned (Fig.3). The southernmost building is a three-bay overhanging gable-roofed front gate (*shanmen*). Inside the first courtyard stands the spectacular Feihong Pagoda (Fig.4), a Buddhist tower with an octagonal ground plan built in *louge* (multistory building)-style, rising up thirteen stories to a total height of 47.95 ms (equal to 15 Ming-period *zhang*). Construction began in 1515, the tenth year of the Zhengde reign period of the Ming dynasty, and took twelve years.

[2] CHAI Zejun, REN Yimin. Hongtong Guangshengsi (Guangsheng Monastery in Hongtong)[M]. Beijing: Wenwu chubanshe, 2006.

○一二

参见：柴泽俊，任毅敏．洪洞广胜寺[M]．北京：文物出版社，2006．

一、广胜上寺

上寺位于一座山峰顶部，整组建筑群依山就势，沿着一条略呈弧形之中轴线分布（图3）。最南为三间悬山顶山门。门内矗立一座飞虹塔（图4），此塔为楼阁式佛塔，八角形平面，十三级，总高47.95米（约合明代15丈），于明代正德十年（1515年）动工，嘉靖六年（1527年）建成。

广胜寺之周遭，山势巍峨，泉水清冽，古柏苍郁，环境幽丽。该寺相传始建于东汉建和元年（147年），原名阿育王塔院、俱卢舍寺，建有阿育王舍利塔一座。北周保定年间（561—565年），高僧正觉曾驻锡于此，睹佛塔颓圮，欲重修而未果。唐大历年间（766—779年），李克瓒、郭子仪等人见寺宇残破，古塔已成土基，遂奏请重建。北宋广胜寺规模尤可观，从时人的诗文中可略窥其意。金代，寺在唐宋基础上略有扩充，由今日水神庙壁画中可见其大致格局——壁画中，广胜上寺呈前塔后阁之布局，山门、塔院门等亦为高阁，规制宏伟，远胜现状（图2）。可惜金贞祐之乱（1213—1217年）中寺庙遭毁。元初，广胜寺及水神庙得以重兴，寺中殿堂斋舍达百楹，庙宇有王宫庄丽之威。然而元大德七年（1303年）的大地震中，全寺尽毁。此后又重建，现存广胜寺上、下寺及水神庙诸殿宇多为此番重建之元构。有明一代，除补葺殿宇外，最重要之工程是飞虹塔的重建。○一二

图2 水神庙明应王殿元代壁画中所见古广胜上寺（来源：贾珺摄）

008

图4 飞虹塔外观（来源：贾珺摄）

图3 广胜上寺远眺（来源：贾珺摄）

Fig.2 Ancient Upper Guangsheng Monastery seen in the Yuan Dynasty murals in the Mingying Wang Hall of Water God's Temple. Source: Photographed by JIA Jun

Fig.3 Overlooking the Upper Guangsheng Monastery. Source: Photographed by JIA Jun

Fig.4 Feihong Pagoda. Source: Photographed by JIA Jun

The pagoda body is made of grey bricks. Eaves and *pingzuo* (here referring to exterior balconies) mimic wooden construction and have bracket sets (*dougong*) and balustrades. The walls are embellished with colorful glazed-tile decorations and appear bright and brilliant (Fig.5). In 1622 (the second year of the Tianqi reign period), a subsidiary structure (*fujie*) surrounding the ground floor was attached to the core building. Flying rafters and eaves rafters of the new corridor are decorated with eave-edge plate tiles on the outer edges. The pagoda is hollow with a narrow staircase inside. A large Buddha statue is installed at the ground floor under a sophisticated and complex caisson ceiling (*zaojing*). As a unique cxamplc among Chincsc pagodas, Fcihong Pagoda has always attracted attcntion—in 1958, the State Post Bureau issued a special stamp of this spectacular building.

North of Feihong Pagoda stands Mituo (Amitābha) Hall, a five-bay hip-gable roofed structure housing an eight-rafter wooden sutra cabinet (*jingchu*) where the Jin-period Zhaocheng Tripitaka was originally stored—Guangsheng Monastery became famous when precious Buddhist scriptures were rediscovered in 1933. One-bay passage doors are attached on both sides of Amitābha Hall. The next hall along the central axis is the Treasure Hall of the Great Hero (Daxiongbaodian), a five-bay structure with overhanging gable roof that was rebuilt in 1309 (the second year of the Zhida reign period of the Yuan dynasty) and repaired in 1452 (the third year of the Jingtai reign period of the Ming dynasty). On the east and west sides of the courtyard, in front of the Treasure Hall, are seven-bay auxiliary buildings. The last hall, located in the third courtyard, is Pilu (Vairocana) Hall, dedicated to Vairocana Buddha, also known as "the God of Gods" (*Tianzhongtian*, the Chinese translation of the Sanskrit term *devatideva*), a five-bay hip-roofed structure framed by a five-bay east-side building and a seven-bay west-side building.

2. Lower Guangsheng Monastery

Lower Guangsheng Monastery is situated next to Huo Spring, and consists of two courtyards. A three-bay hip-gable roofed front gate (*shanmen*), also known as Hall of Heavenly Kings (Tianwangdian), leads into the first courtyard (Fig.6). Since waist eaves are installed for rainwater to drain off, the hall appears to have two sets of eaves along the front and rear façades (but only one set of eaves on the broadsides). The slope of the front waist eaves is elongated. Inside the courtyard stands Mituo (Amitābha) Hall, a five-bay structure with overhanging gable roof. A plague from 1892 (the eighteenth year of the Guangxu reign period of the Qing dynasty) is installed below the eaves, inscribed with the four characters "*bao* (precious) *fa* (raft) *jin* (golden) *sheng* (rope)". To the east and west of the hall are Qing-period drum and bell towers, which are two-storied structures of exquisite appearance with arched passageway on the ground floor and cross-ridge hip-gable roof (*shizi xieshan*

塔身主要用青砖砌筑，檐部和平坐模仿木构雕刻斗栱、栏杆，墙面上带有大量琉璃装饰，极其鲜亮（图5）。天启二年（1622年）在塔底层四周加建副阶回廊，回廊飞椽和檐椽端部装有珍贵的琉璃椽头盘子。塔内部中空，设有狭窄的踏道，底层供奉大型佛像，顶部有复杂的藻井雕饰。此塔是中国古塔空见精品，1958年即登上国家邮票。

塔北为前殿弥陀殿，五间歇山建筑，殿内有木制经橱八架，原为存放金版大藏经即著名的『赵城藏』之所在——1930年代，广胜寺曾因『赵城藏』而名满天下。殿左右各有一间过门。中殿即大雄宝殿，五间悬山建筑，元代至大二年（1309年）重建，明代景泰三年（1452年）重修，院东西两侧各有七间配殿。最后一进院北为后殿毗卢殿，又名『天中天』，五间庑殿建筑，东有五间配殿，西为七间配殿。

二、广胜下寺

下寺依临霍泉，主院分为前后两进。前设三间山门（图6），兼作天王殿，三间歇山建筑，前后出单坡腰檐（垂花雨搭），故而前后看若重檐，两山看为单檐，造型别致，南引长长的坡道。门内前殿名为弥陀殿，五间悬山建筑，檐下悬光绪十八年（1892年）所书『宝筏金绳』匾。前殿东西两侧有清代所建的钟鼓楼，均分为上下两层，底层设拱门过道，上层采用十字歇山屋顶，造型精巧。后殿又名大佛殿，殿内供奉三世佛，东西各设配殿，殿内四壁原本满绘精美壁画，于1928年被卖，现藏于美国堪萨

图6 广胜下寺山门（来源：贾珺摄）

图5 飞虹塔琉璃雕饰（来源：贾珺摄）

Fig.5 The colorful glazed-tile decorations of Feihong Pagoda. Source: Photographed by JIA Jun
Fig.6 *shanmen* of Lower Guangsheng Monastery. Source: Photographed by JIA Jun

wuding). The rear hall in the second courtyard is known as Great Buddha Hall and houses the three Buddhas of the Past, Present, and Future. East- and west-side buildings stand on both sides of the courtyard. The murals on the walls of Great Buddha Hall are some of the best known in Chinese art history, perhaps because they were sold in 1928 and are now in the possession of the Nelson-Atkins Museum in Kansas City, United States. Additionally, a three-bay ear building (*duodian*) flanks the western side of Great Buddha Hall. Noteworthy is the use of enlarged cantilevers (*da'ang*) in the timer frameworks of the Yuan-period halls of both Upper and Lower Guangsheng Monastery. LIN Huiyin and LIANG Sicheng wrote:

"Among the many relics that we saw in China over the past years, this structure is still the first of its kind. It is particularly important because traditional Japanese buildings especially the early-period monasteries like Asukadera use enlarged cantilevers (*ang*) and apply the same construction method that we can see here. This extant example confirms what we have long suspected—that the corresponding Japanese method derives in fact from a Chinese pre-Song architectural principle and is not at all developed by its own efforts. The principle simply failed to be handed down over successive generations in China or is just seldom seen. At the same time, we believe that Guangsheng Monastery construction is much more significant in terms of actual buildings as we originally imagined."[3]

3. Water God's Temple

The Water God's Temple is an independent courtyard situated on the western side of Lower Guangsheng Monastery, sharing its east wall with the monastery, where the spirit of Huo Spring (Mingying Wang) is worshipped. A three-bay stage with overhanging gable roof stands south of the hall and functions as the entrance gate to the temple. The south wall has an arched brick gate, inscribed with the four characters of "*shen* (spirit) *wei* (power) *zheng ya* (suppress)". Opposite the gate stand two 3-m tall statues of the Water God. Brick carvings decorate the partition wall installed in the central bay. The visitors must pass another gate (*shanmen*) north of the stage leading to the main hall known as Mingying Wang Hall, which is a three-bay hip-gable roofed structure surrounded by a corridor. Rebuilt in 1319 (the sixth year of the Yanyu reign period of the Yuan dynasty), the walls are painted all around with exquisite murals depicting theatrical scenes and historical events, as well as telling stories about offerings for rain. With their fine brushwork and vivid strokes, they are outstanding examples of Yuan wall painting (Fig.7, Fig.8).

③　LIN Whei-Yin, LIANG Ssu-ch'eng. A survey record of the traditional architecture in Fen region, Shanxi province (Jin Fen gujianzhu yucha jilüe)[J]. Journal of the Society for Research in Chinese Architecture (Zhongguo yingzao xueshe huikan), 1935, 5(3).

斯城纳尔逊艺术馆及纽约大都会博物馆。后殿西端另存三间西朵殿。广胜上、下寺各元代木构殿宇中，大昂的运用为极突出之特征。林徽因、梁思成写道：「这种构架，在我们历年国内各地所见许多的遗物中，这还是第一个例。尤其重要的，是因日本的古建筑，尤其是飞鸟、灵乐等初期的遗构，都是用极大的昂，结构法与此相类，这个实例乃大可佐证建筑家早就怀疑的问题，这问题便是日本这种结构法，是直接承受中国以前建筑规制，并非自创，而此种规制，在中国后代反倒失传或罕见。同时我们相信广胜寺各构，在建筑遗物实例中的重要，远超过于我们起初所想像的。」[2]

三、水神庙

水神庙为下寺西侧一组独立的庭院，主要祭祀霍泉之神。最南是一座戏台，兼作前山门，三间悬山建筑，南设砖砌拱门，上雕『神威镇压』四字额，门内对称竖立两尊3米高的天神像，中间隔墙上有精致的砖雕。戏台北为山门，门内为正殿明应王殿，三开间周围廊歇山顶建筑。重建于元代延祐六年（1319年），四壁绘满壁画，主要表现祈神降雨以及历史故事、戏剧场景，笔法细腻生动，是元代壁画的上乘之作（图7、图8）。

（二）　林徽因，梁思成. 晋汾古建筑预查纪略[J]. 中国营造学社汇刊，1935，5（3）.

图8　水神庙明应王殿壁画之二（来源：贾珺摄）

图7　水神庙明应王殿壁画之一（来源：贾珺摄）

Fig.7　The mural in the Mingying Wang Hall of Water God's Temple (I). Source: Photographed by JIA Jun

Fig.8　The mural in the Mingying Wang Hall of Water God's Temple (II). Source: Photographed by JIA Jun

In 2013 and 2014, the School of Architecture at Tsinghua University organized two field trips to Guangsheng Monastery, where undergraduate students (Classes of 2010 and 2011) surveyed and mapped the historical buildings under leadership of Tsinghua professors (WANG Guixiang, LIU Chang, JIA Jun, QING Feng, and LI Luke) and graduate students (LIU Mengyu, BAO Aidi, LI Qinyuan, ZHAO Bo, XU Yang, LI Minhao, WANG Xichen, XU Teng, HUANG Wenhao, etc.). Next to traditional hand measuring tools, we used total stations, 3D laser scanners, and photogrammetry technology. Additionally, we documented the inscriptions on stone steles and wooden plaques. This was the most comprehensive surveying and mapping of the architecture at Guangsheng Monastery to date. The accurate data we collected will facilitate monument preservation and allow future research at a broad and a detailed level.

JIA Jun, WANG Nan

Architecture History and Historic Preservation Research Institute, School of Architecture, Tsinghua University

Translated by Alexandra Harrer

2013年、2014年，清华大学建筑学院师生先后两次针对广胜寺展开测绘，带队教师为王贵祥、刘畅、贾珺、青锋、李路珂，参加的研究生包括刘梦雨、包媛迪、李沁园、赵波、徐扬、李旻昊、王曦晨、徐腾、黄文镐等，参加的本科生主要为2010级和2011级的同学。本次测绘综合运用了三维激光扫描、近景摄影测量、全站仪以及各种手工工具，并对碑刻、题记作了全面的记录，是有史以来针对广胜寺最全面的一次测绘调查，成果可为未来进一步的建筑历史研究和古建筑保护提供详实的资料。

清华大学建筑学院 建筑历史与文物保护研究所

贾珺 王南

图

版

Figures

广胜上寺

Upper Guangsheng
Monastery

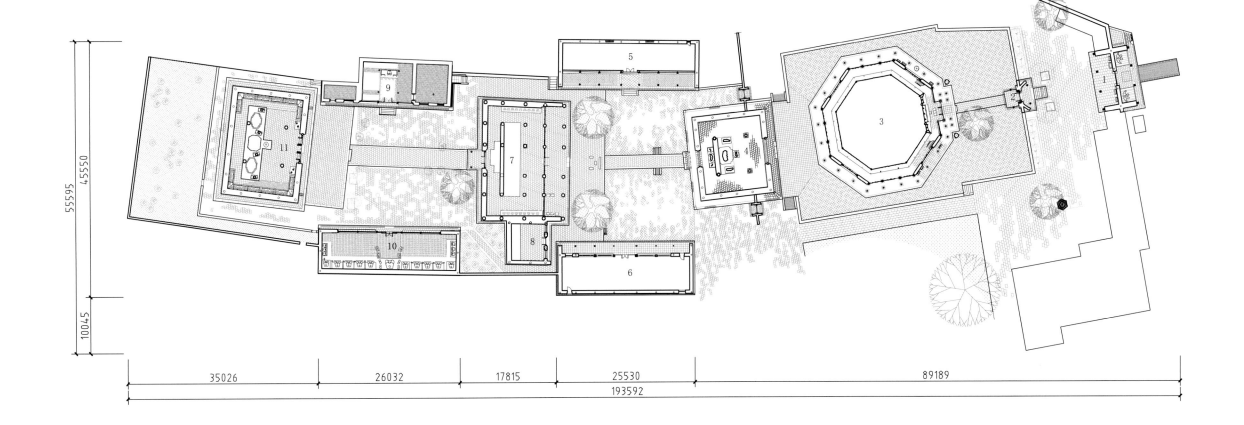

1　山门　　　　7　大雄宝殿
2　塔院小门　　8　西垛殿
3　飞虹塔　　　9　观音殿
4　弥陀殿　　　10　地藏殿
5　东配殿　　　11　毗卢殿
6　西配殿

广胜上寺总平面图
Overall Layout of the Upper Guangsheng Monastery

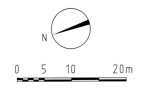

0　5　10　　　20m

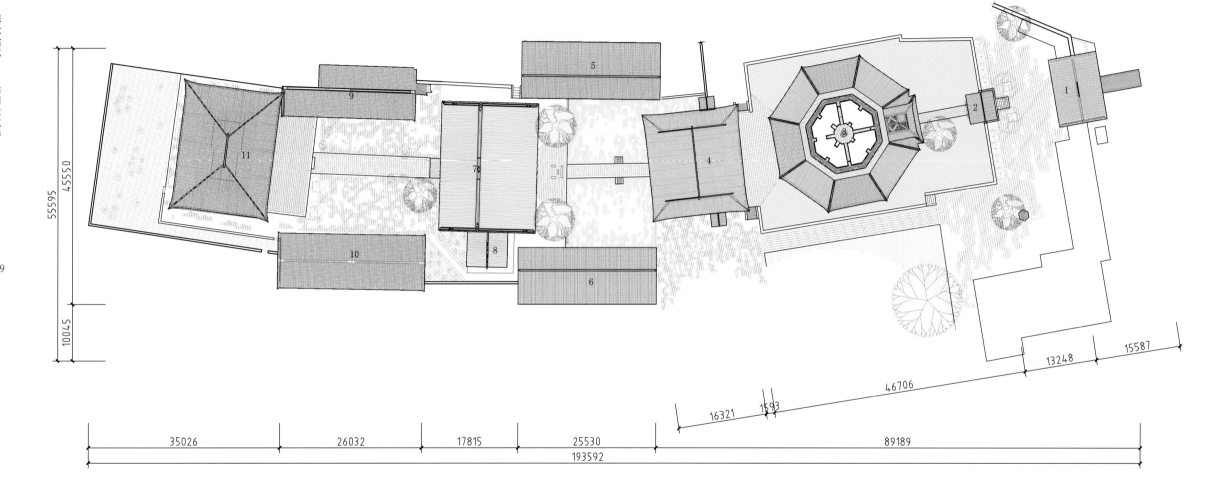

1　山门　　7　大雄宝殿
2　塔院小门　8　西垛殿
3　飞虹塔　　9　观音殿
4　弥陀殿　　10　地藏殿
5　东配殿　　11　毗卢殿
6　西配殿

广胜上寺屋顶总平面图
Overall Topview of Roofs, Upper Guangsheng Monastery

0　5　10　　20m

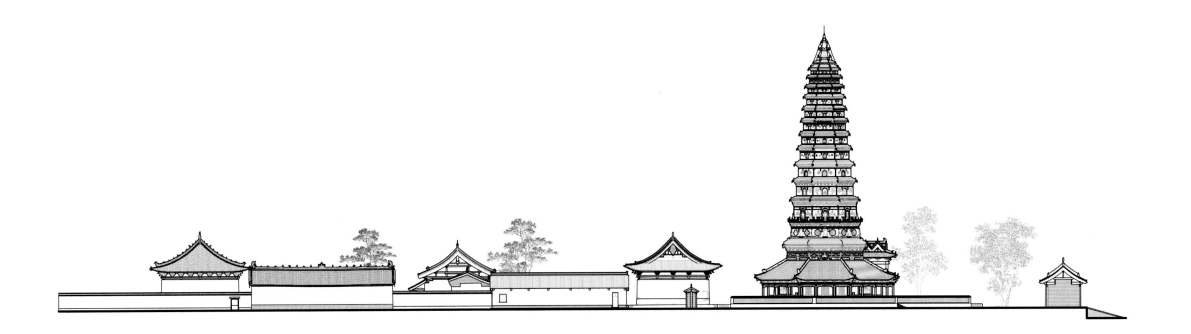

广胜上寺西立面展开图
Unfolded Drawing of the West Elevation, Upper Guangsheng Monastery

0　5　10　　　20m

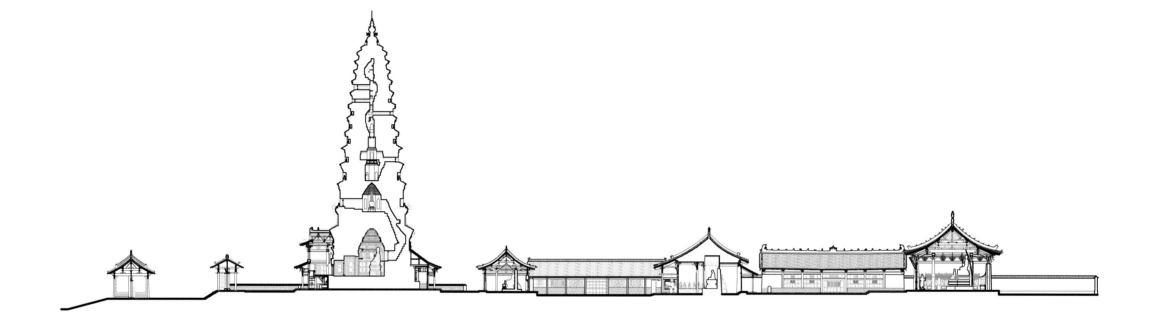

广胜上寺纵剖面图
Longitudinal Section of the Upper Guangsheng Monastery

0　5　10　　　20m

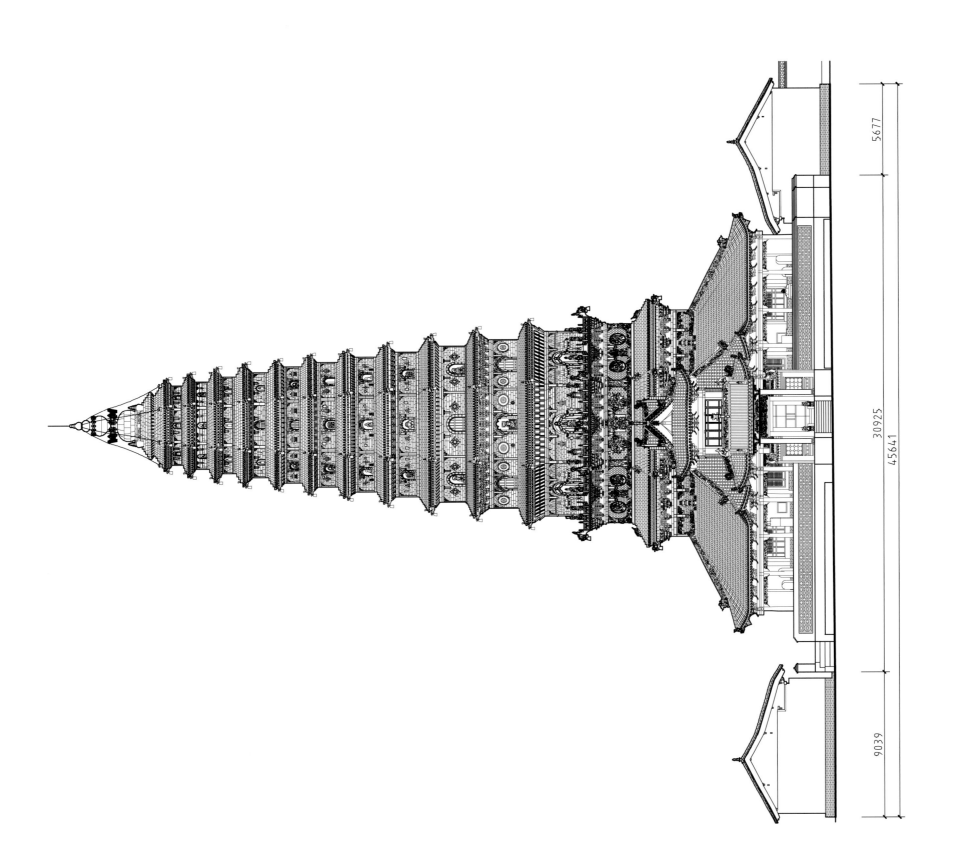

广胜上寺前院横剖面图

Cross Cross Section of the Forecourt, Upper Guangsheng Monastery

5677

30925

45641

9039

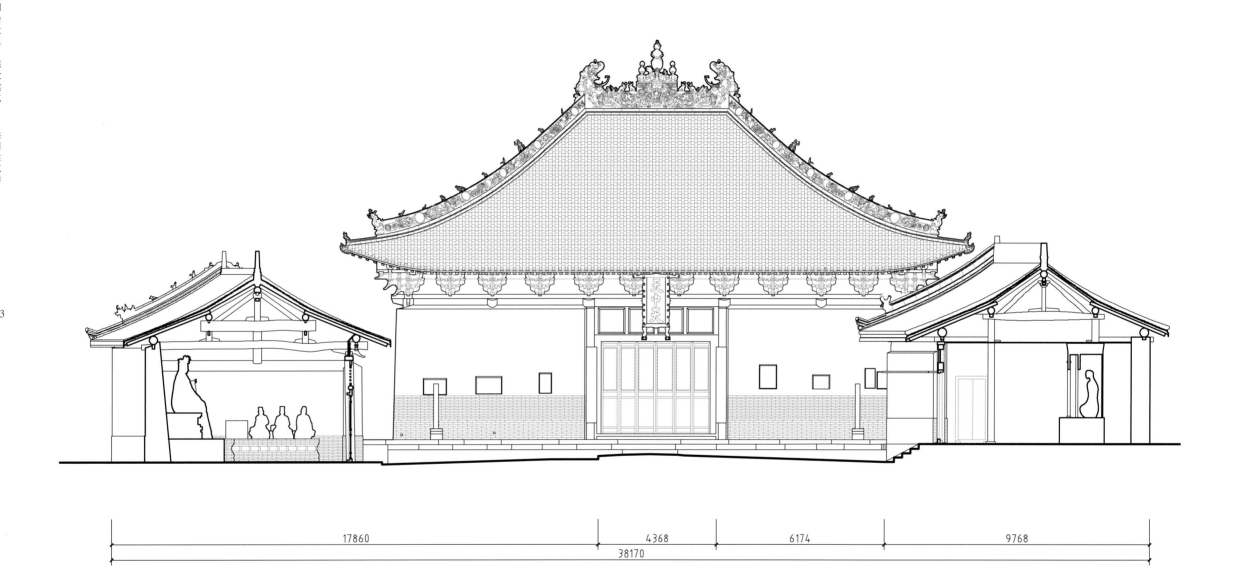

| 17860 | 4368 | 6174 | 9768 |

38170

广胜上寺后院横剖面图
Cross Cross Section of the Backcourt, Upper Guangsheng Monastery

0　1　　　3m

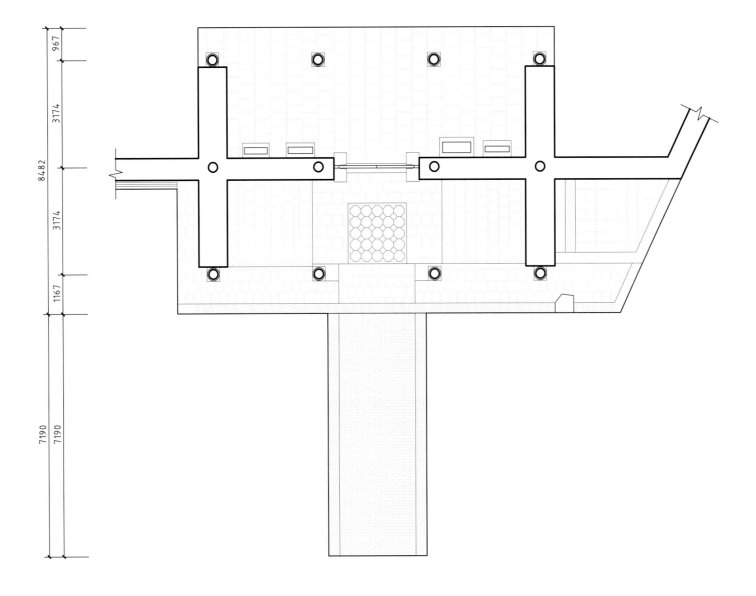

广胜上寺山门平面图
Floor Plan of the Front Gate, Upper Guangsheng Monastery

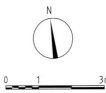

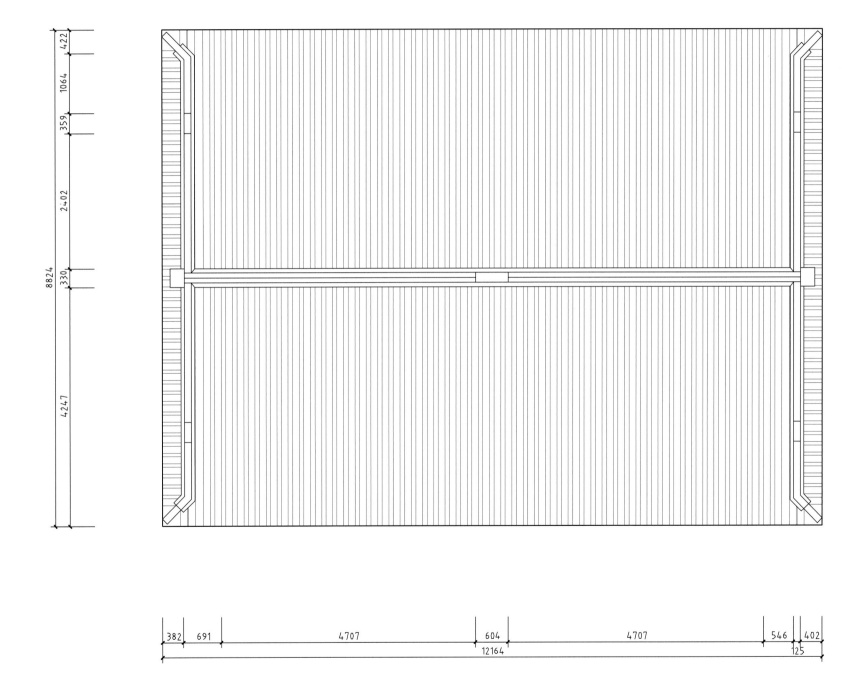

广胜上寺山门屋顶平面图
Top View of the Roof of the Front Gate, the Upper Guangsheng Monastery

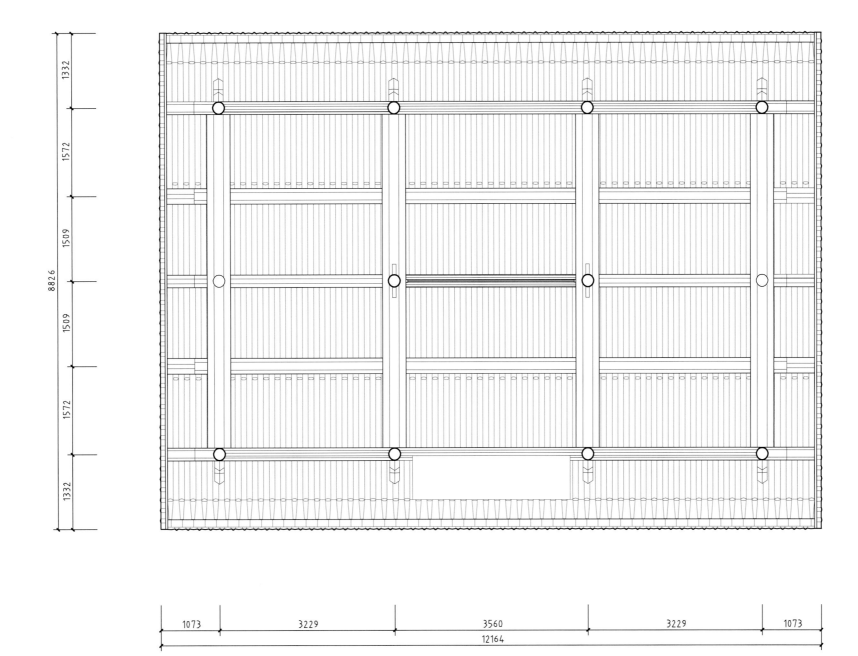

广胜上寺山门梁架仰视平面图
Bottom View of the Truss of the Front Gate, Upper Guangsheng Monastery

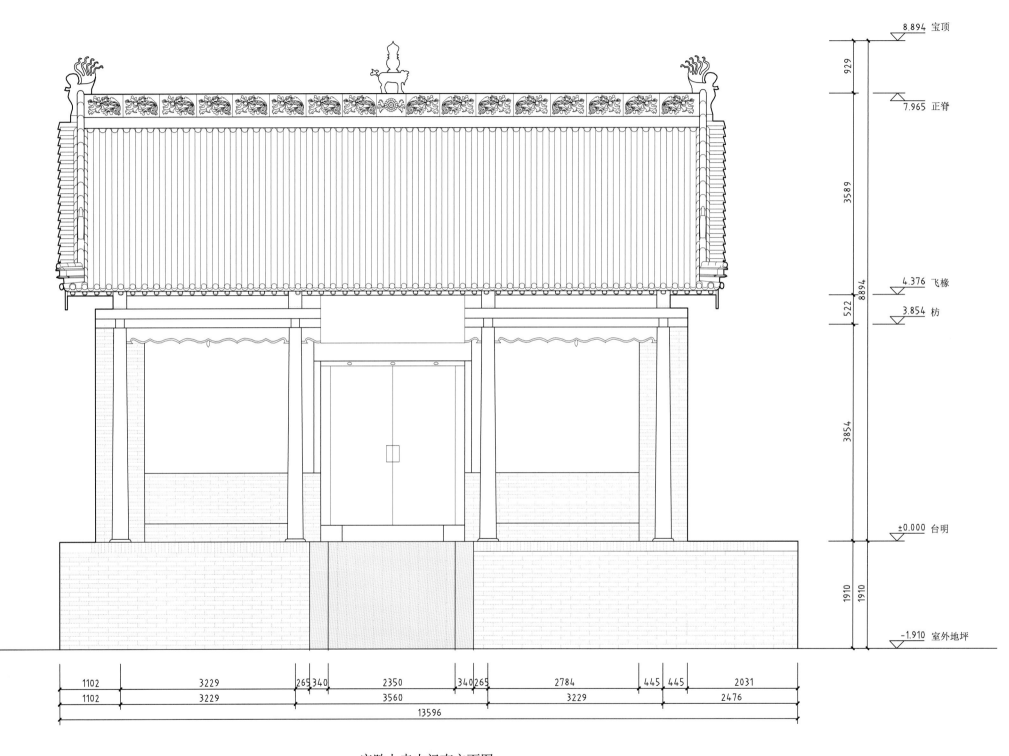

8.894 宝顶

7.965 正脊

4.376 飞椽

3.854 枋

±0.000 台明

-1.910 室外地坪

929

3589

522

8894

3854

1910

1910

1102　3229　265 340　2350　340 265　2784　445 445　2031

1102　3229　3560　3229　2476

13596

广胜上寺山门南立面图
South Elevation of the Front Gate, Upper Guangsheng Monastery

0　0.5　1　2m

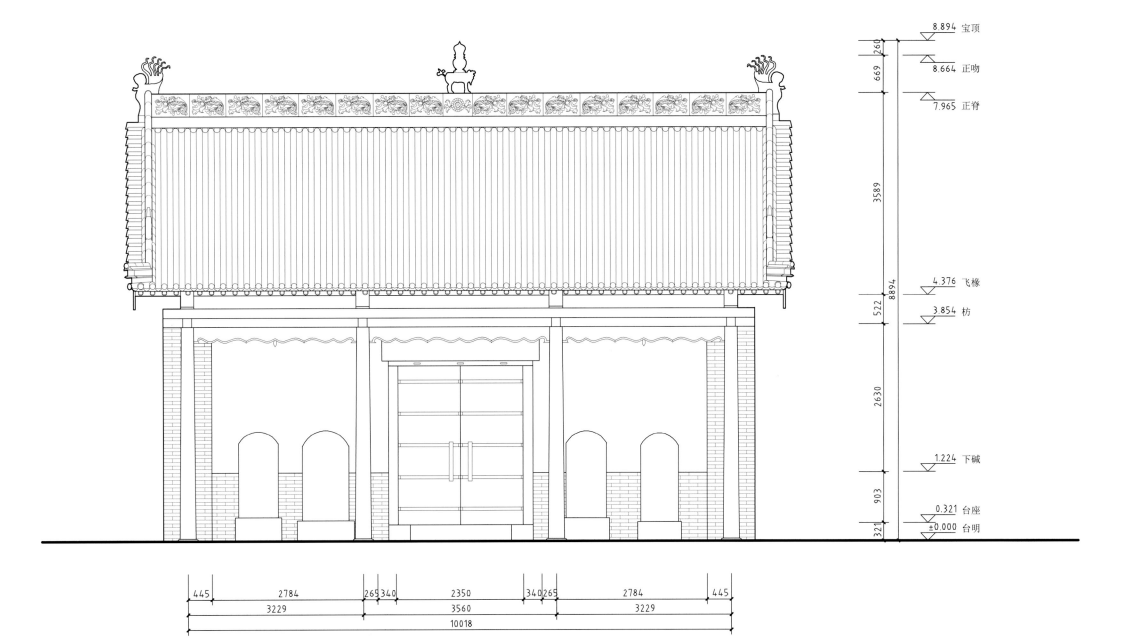

8.894 宝顶

8.664 正吻

7.965 正脊

4.376 飞椽

3.854 枋

1.224 下碱

0.321 台座

±0.000 台明

260

669

3589

522

2630

903

321

8894

445　2784　265 340　2350　340 265　2784　445

3229　3560　3229

10018

广胜上寺山门北立面图
North Elevation of the Front Gate, Upper Guangsheng Monastery

0　0.5　1　2m

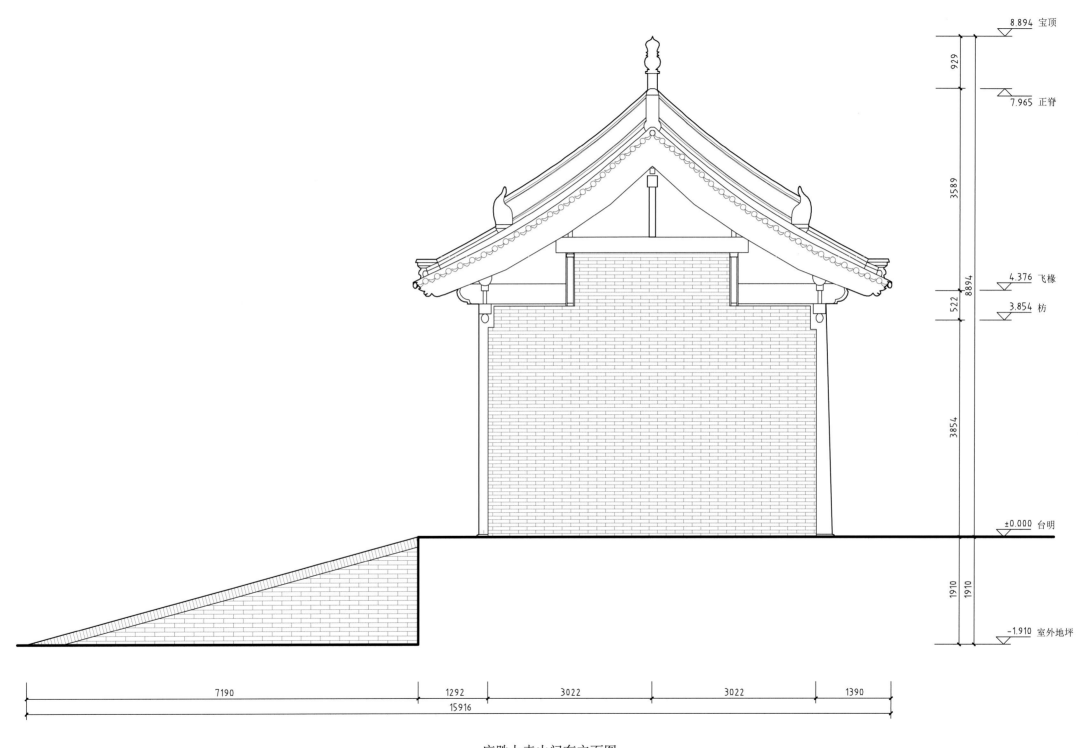

8.894 宝顶

929

7.965 正脊

3589

4.376 飞椽

8894

522

3.854 枋

3854

±0.000 台明

1910

1910

-1.910 室外地坪

7190　1292　3022　3022　1390

15916

广胜上寺山门东立面图
East Elevation of the Front Gate, Upper Guangsheng Monastery

0　0.5　1　2m

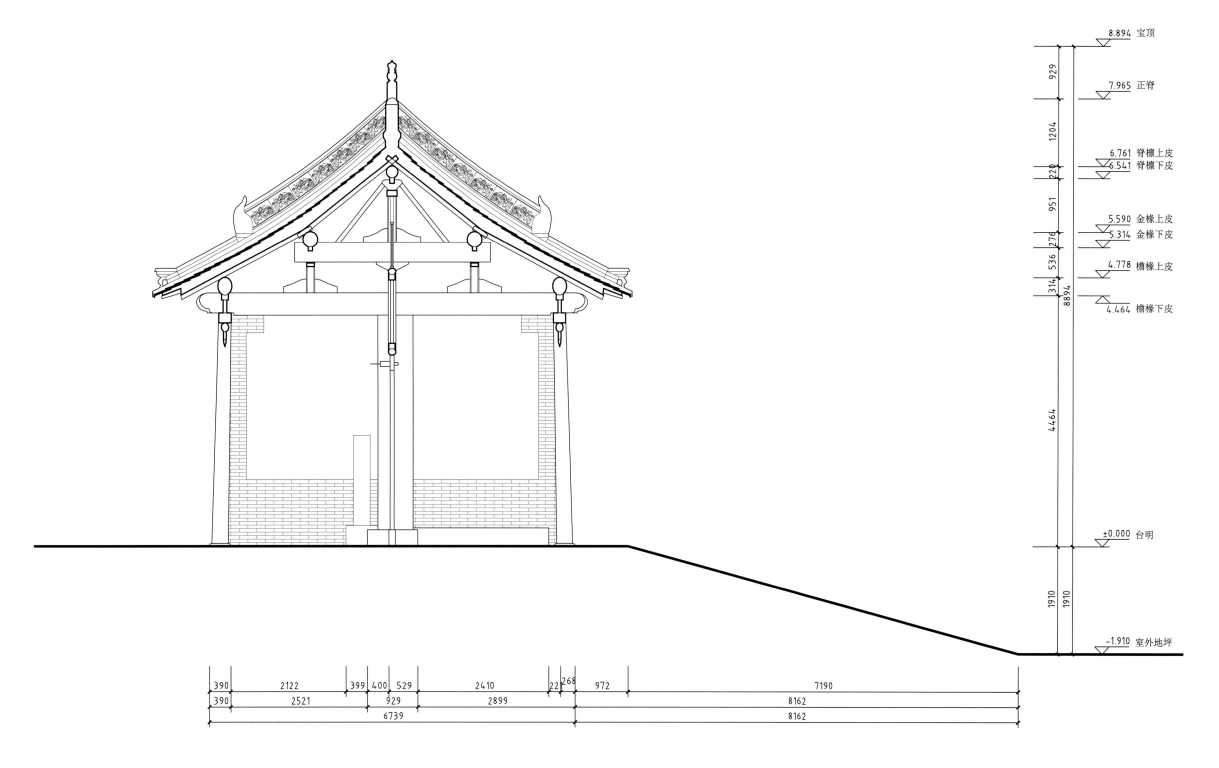

8.894 宝顶

7.965 正脊

6.761 脊檩上皮
6.541 脊檩下皮

5.590 金檩上皮
5.314 金檩下皮

4.778 檐檩上皮

4.464 檐檩下皮

929

1204

220

951

276

536

314

8894

±0.000 台明

4464

1910

1910

−1.910 室外地坪

390　2122　399 400 529　2410　22 268　972　7190

390　2521　929　2899　8162

6739　8162

广胜上寺山门明间横剖面图

Cross Section of the Central Bay of the Front Gate, Upper Guangsheng Monastery

0　0.5　1　2m

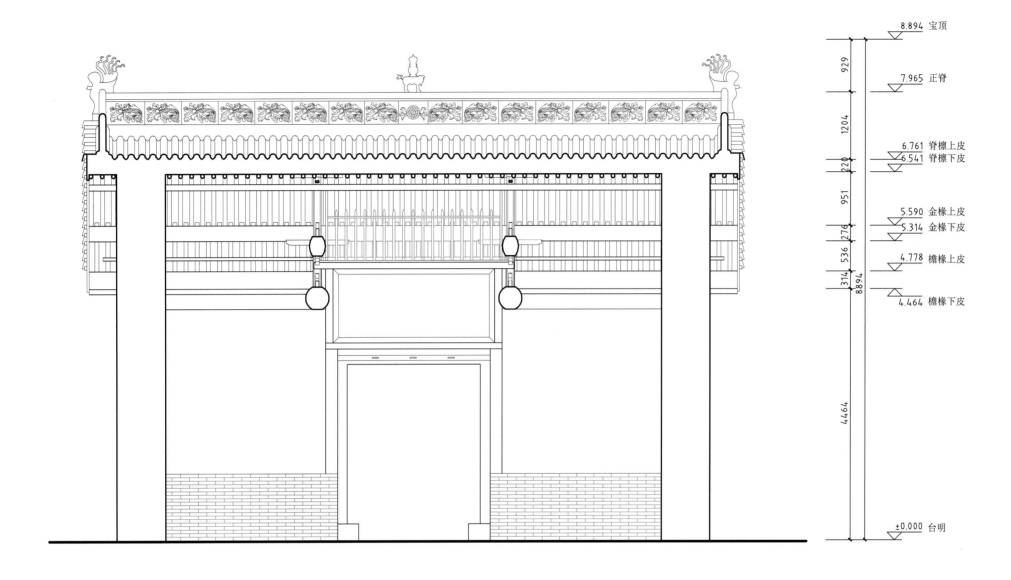

8.894 宝顶

7.965 正脊

6.761 脊檩上皮
6.541 脊檩下皮

5.590 金檩上皮

5.314 金檩下皮

4.778 檐檩上皮

4.464 檐檩下皮

±0.000 台明

929
1204
220
951
276
536
314
8894

4464

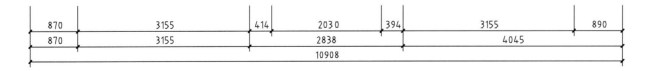

| 870 | 3155 | 414 | 2030 | 394 | 3155 | 890 |

| 870 | 3155 | 2838 | 4045 |

10908

广胜上寺山门纵剖面图
Longitudinal Section of the Front Gate, Upper Guangsheng Monastery

0 0.5 1 2m

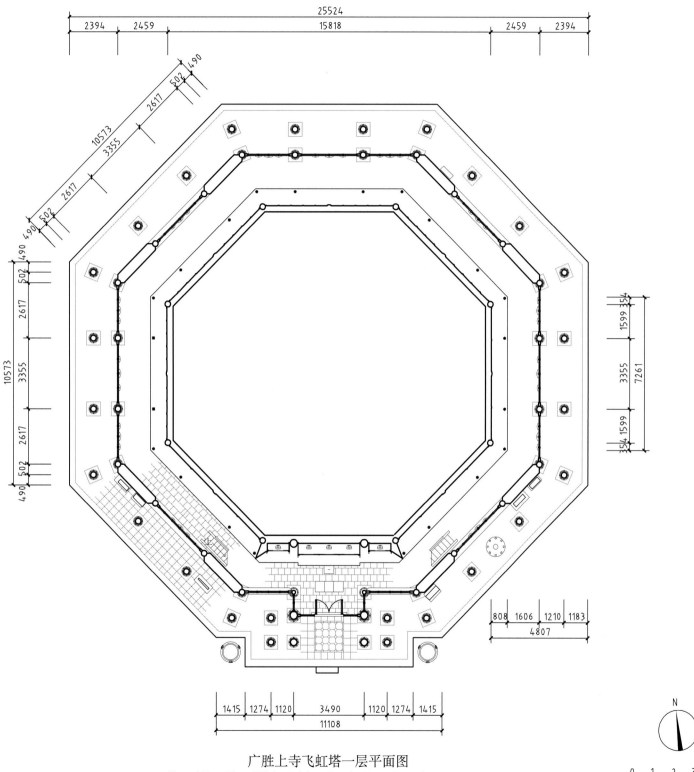

广胜上寺飞虹塔一层平面图
Ground Floor Plan of Flying Rainbow Pagoda, Upper Guangsheng Monastery

N

0 1 2 3 4 5m

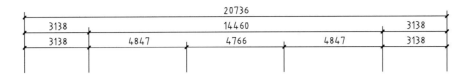

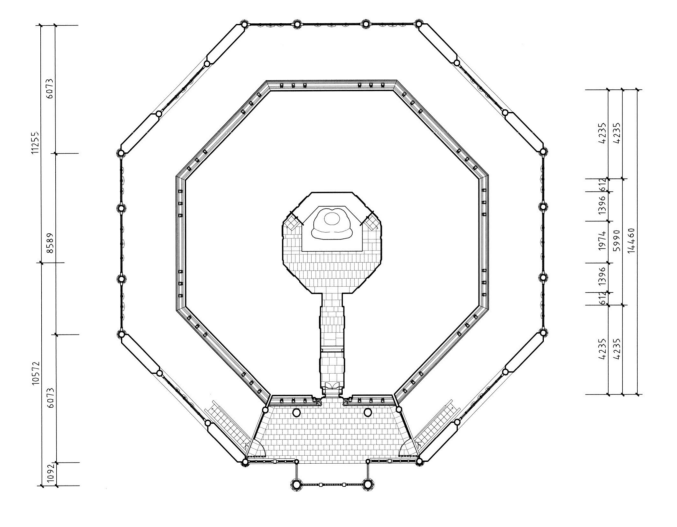

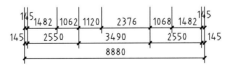

广胜上寺飞虹塔二层平面图
Second Floor Plan of Flying Rainbow Pagoda, Upper Guangsheng Monastery

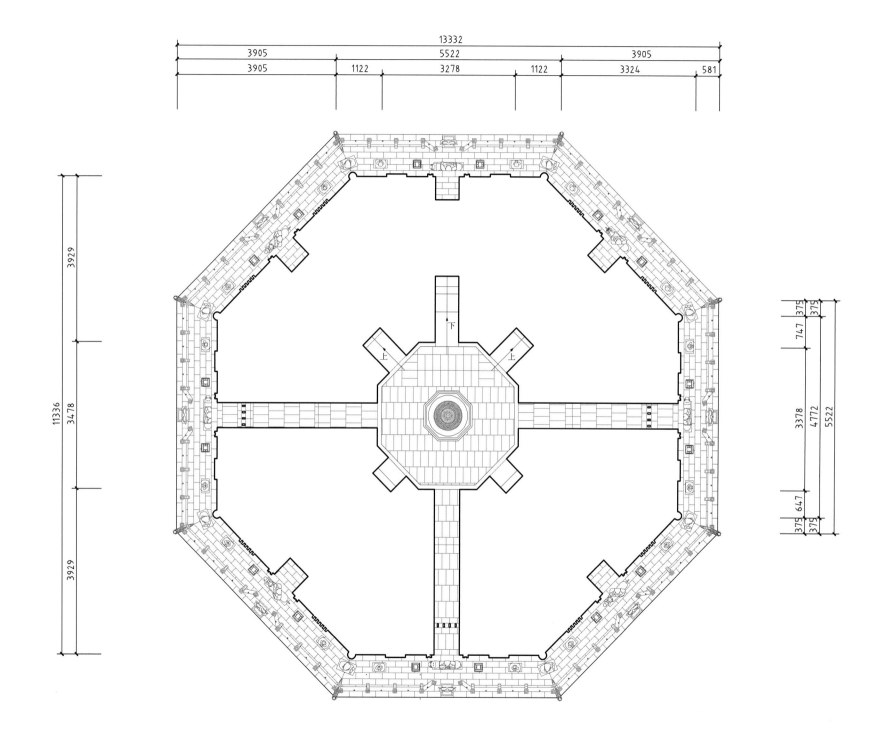

广胜上寺飞虹塔三层平面图
Third Floor Plan of Flying Rainbow Pagoda, Upper Guangsheng Monastery

0 1 2m

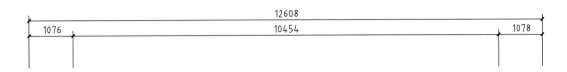
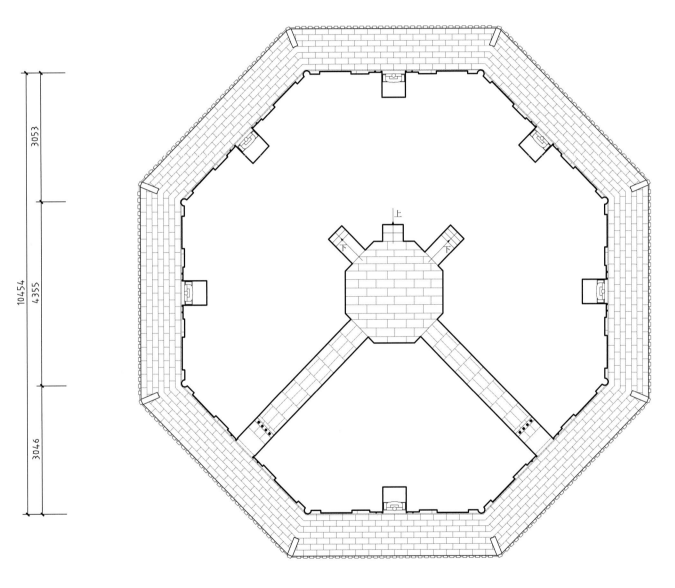

广胜上寺飞虹塔四层平面图
Fourth Floor Plan of Flying Rainbow Pagoda, Upper Guangsheng Monastery

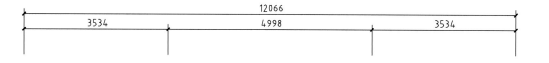

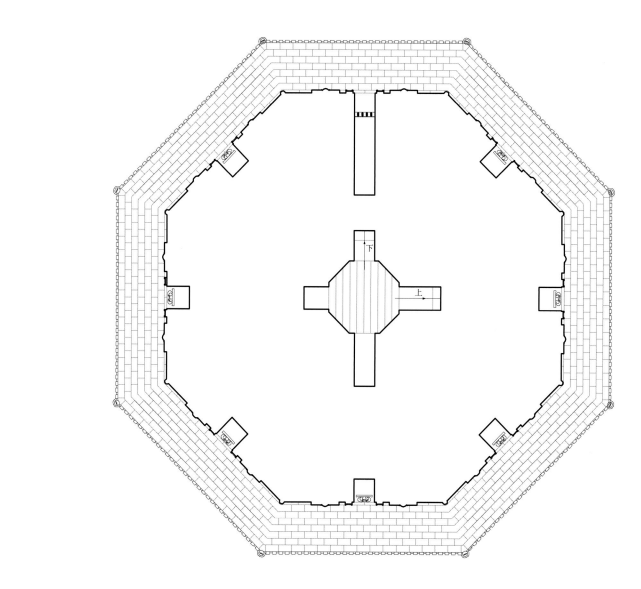

广胜上寺飞虹塔五层平面图
Fifth Floor Plan of Flying Rainbow Pagoda, Upper Guangsheng Monastery

0   1   2m

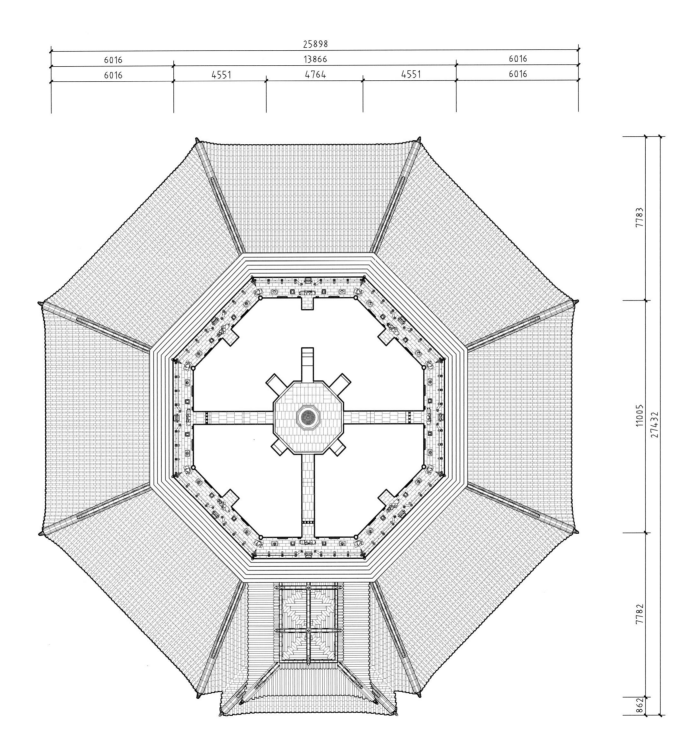

广胜上寺飞虹塔屋顶平面图
Top View of the Roof of Flying Rainbow Pagoda, Upper Guangsheng Monastery

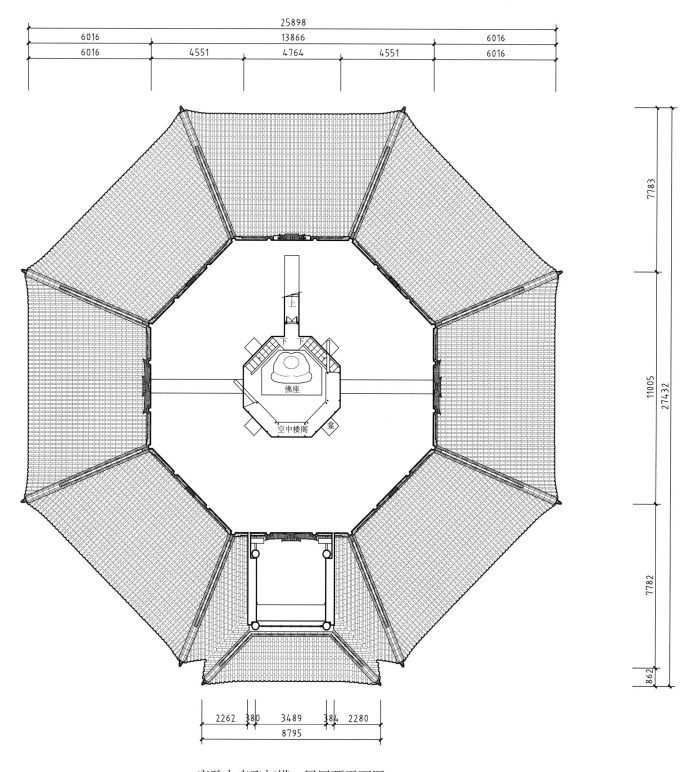

广胜上寺飞虹塔一层屋顶平面图
Top View of the Ground Floor's Roof of Flying Rainbow Pagoda, Upper Guangsheng Monastery

0 1 2 3 4 5m

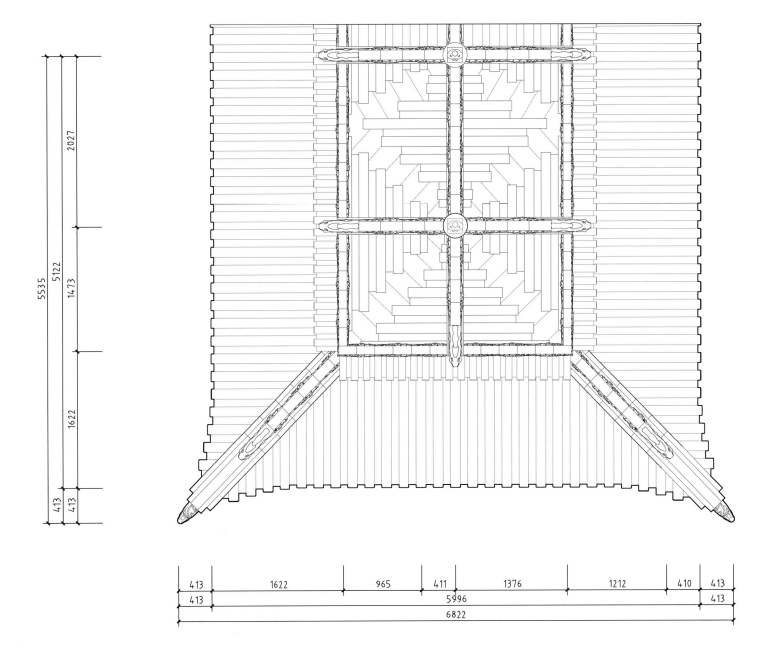

广胜上寺飞虹塔抱厦屋顶平面图
Top View of the Subsidiary Corridor's Roof of Flying Rainbow Pagoda, Upper Guangsheng Monastery

0   0.5   1   1.5m

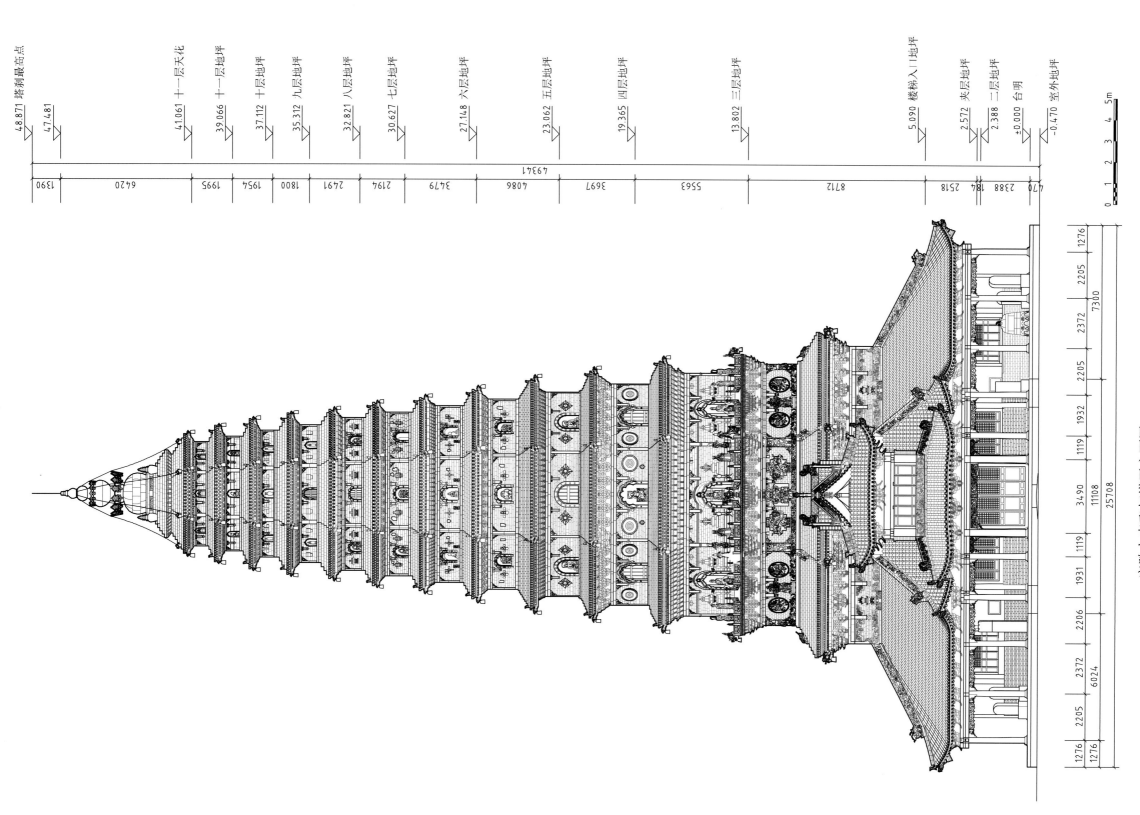

48.871 塔刹最高点
47.481
41.061 十一层天花
39.066 十一层地坪
37.112 十层地坪
35.312 九层地坪
32.821 八层地坪
30.627 七层地坪
27.148 六层地坪
23.062 五层地坪
19.365 四层地坪
13.802 三层地坪
5.090 楼梯入口地坪
2.572 夹层地坪
2.388 二层地坪
±0.000 台明
-0.470 室外地坪

1390
6420
1995
1954
1800
2491
2194
3479
4086
3697
5563
8712
2518 184
2388
470

49341

0 1 2 3 4 5m

1276
2205
2372
2205
1932
1119
3490
1119
1931
2206
2372
2205
1276

7300
11108
25708
6024
1276

广胜上寺飞虹塔正立面图
Front Elevation of Flying Rainbow Pagoda, Upper Guangsheng Monastery

041

广胜上寺飞虹塔侧立面图
Side Elevation of Flying Rainbow Pagoda, Upper Guangsheng Monastery

48.871 塔刹最高点
47.481
41.061 十一层天花
39.066 十一层地坪
37.112 十层地坪
35.312 九层地坪
32.821 八层地坪
30.627 七层地坪
27.148 六层地坪
23.062 五层地坪
19.365 四层地坪
13.802 三层地坪
5.090 楼梯入口地坪
2.572 夹层地坪
2.388 二层地坪
±0.000 台明
-0.470 室外地坪

1390
6420
1995
1954
1800
2491
2194
3479
4086
3697
5563
8712
2518
184
2388
470

49341

0 1 2 3 4 5m

广胜上寺飞虹塔背立面图
Back Elevation of Flying Rainbow Pagoda, Upper Guangsheng Monastery

48.871 塔刹最高点
47.481
41.061 十一层天花
39.066 十一层地坪
37.112 十层地坪
35.312 九层地坪
32.821 八层地坪
30.627 七层地坪
27.148 六层地坪
23.062 五层地坪
19.365 四层地坪
13.802 三层地坪
5.090 楼梯入口地坪
2.572 夹层地坪
2.388 二层地坪
±0.000 台明
-0.470 室外地坪

1390
6420
1995
1954
1800
2491
2194
3479
4086
3697
5563
8712
2518
2388 184 2518
470

49341

1276
2205
2372
7300
2205
1932
1119
3490
1119
11108
1931
2206
2372
6024
2205
1276
1276
25708

0 1 2 3 4 5m

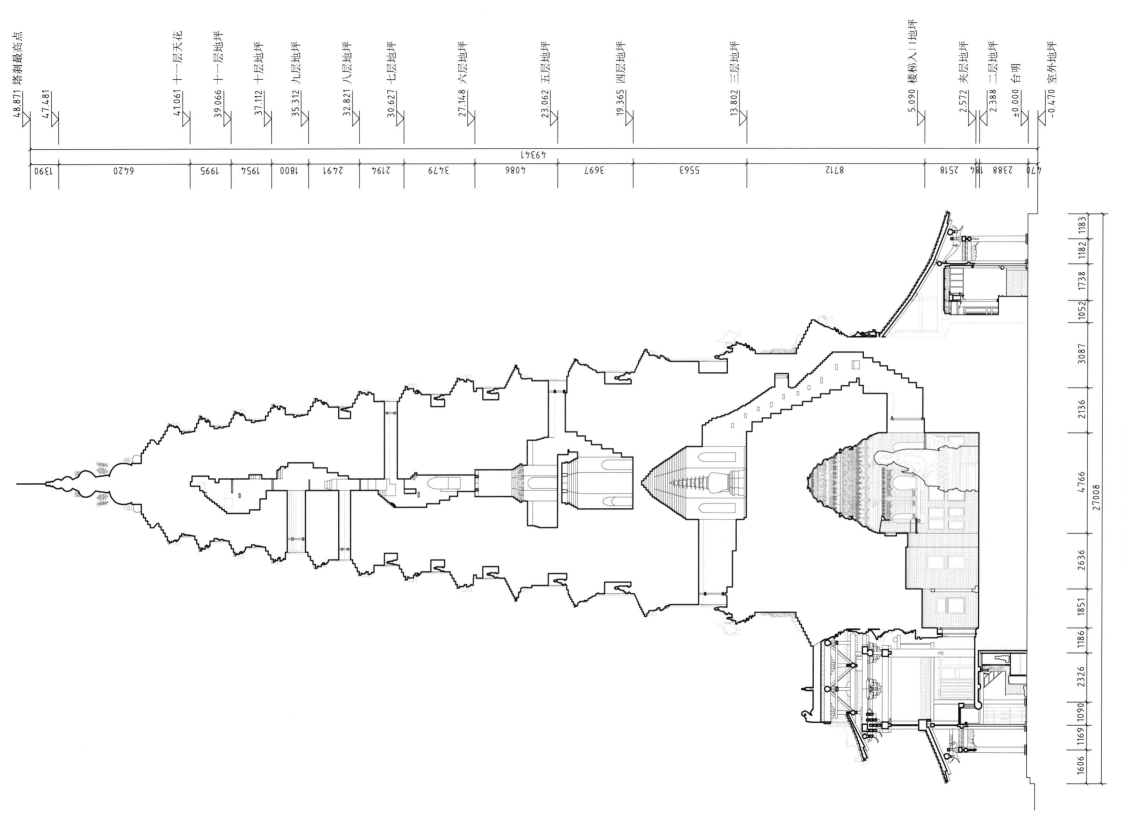

48.871 塔刹最高点
47.481
41.061 十一层天花
39.066 十一层地坪
37.112 十层地坪
35.312 九层地坪
32.821 八层地坪
30.627 七层地坪
27.148 六层地坪
23.062 五层地坪
19.365 四层地坪
13.802 三层地坪
5.090 楼梯入口地坪
2.572 夹层地坪
2.388 二层地坪
±0.000 台明
-0.470 室外地坪

1390
6420
1995
1954
1800
2491
2194
3479
4086
3697
5563
8712
2518 184 2388
470

49341

广胜上寺飞虹塔剖面图
Section of Flying Rainbow Pagoda, Upper Guangsheng Monastery

1183 1182 1738 1052 3087 2136 4766 2636 1851 1186 2326 1090 1169 1606
27008

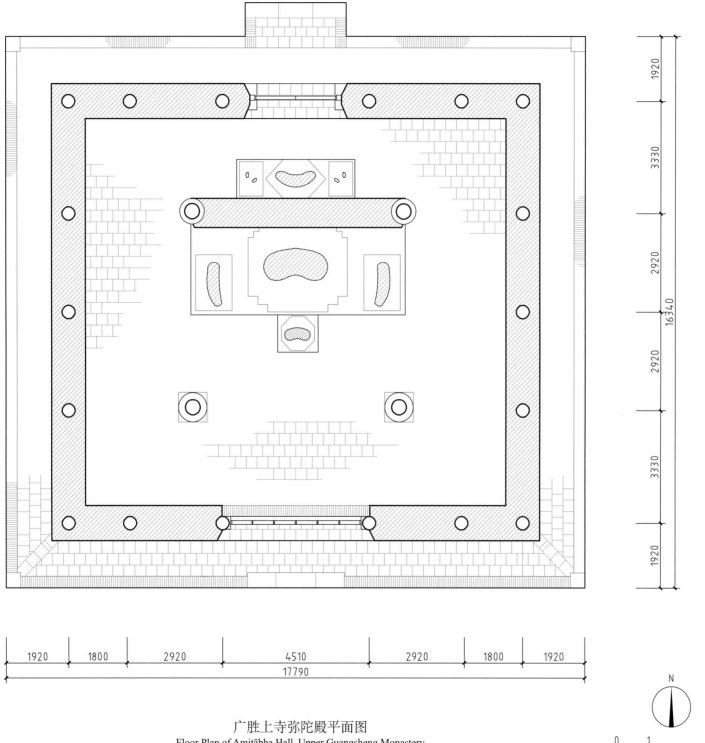

广胜上寺弥陀殿平面图
Floor Plan of Amitābha Hall, Upper Guangsheng Monastery

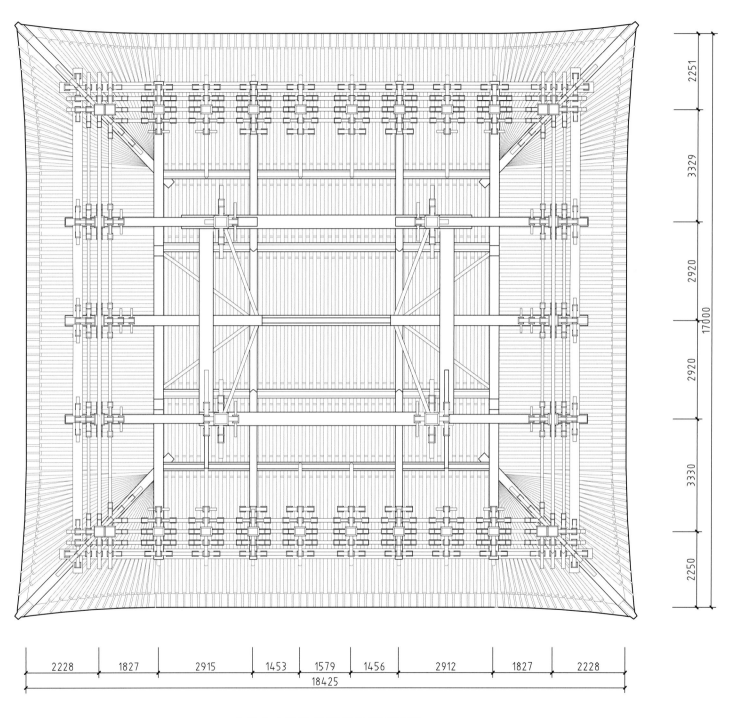

广胜上寺弥陀殿梁架仰视平面图
Bottom View of the Truss of Amitābha Hall, Upper Guangsheng Monastery

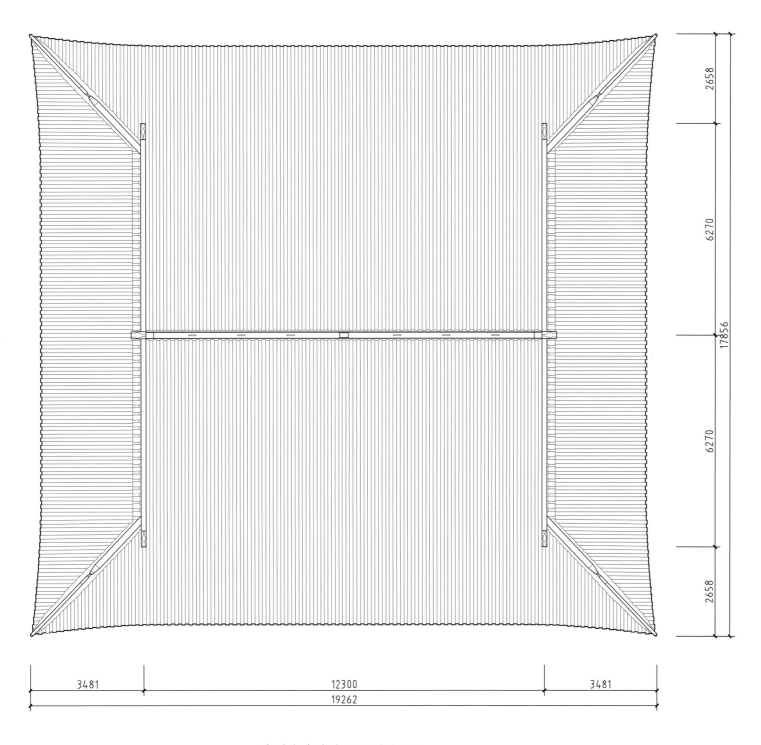

2658

6270

17856

6270

2658

3481

12300

3481

19262

广胜上寺弥陀殿屋顶平面图
Top View of the Roof of Amitābha Hall, Upper Guangsheng Monastery

0　1　　3m

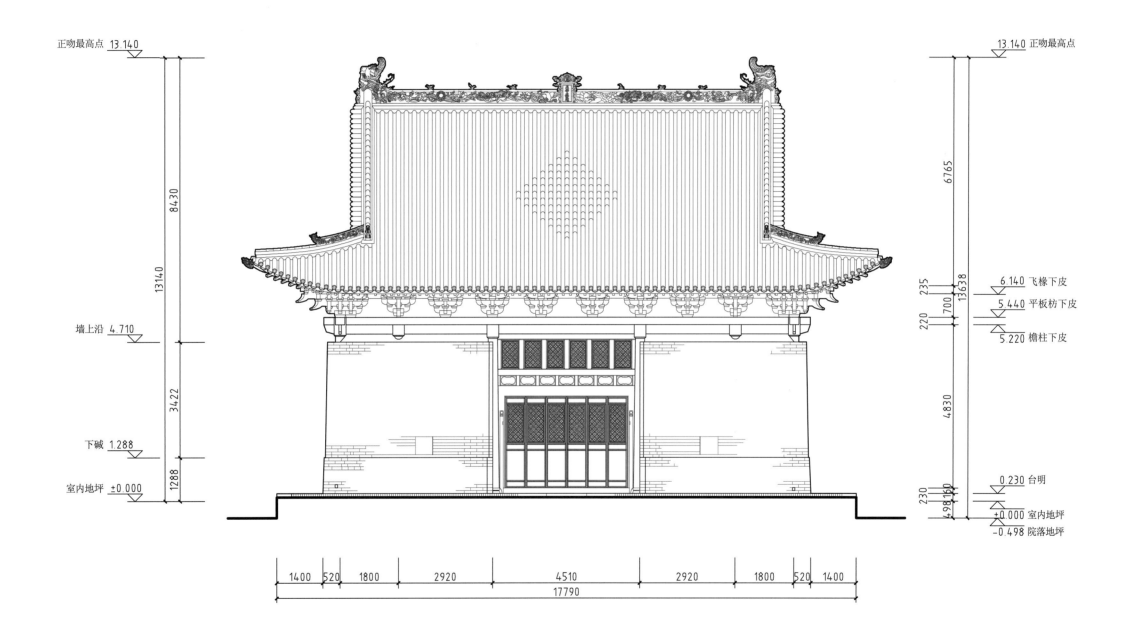

正吻最高点 13.140

墙上沿 4.710

下碱 1.288

室内地坪 ±0.000

8430

13140

3422

1288

6765

235
220
700
13638

6.140 飞椽下皮
5.440 平板枋下皮
5.220 檐柱下皮

4830

230
150
4.98

0.230 台明
+0.000 室内地坪
-0.498 院落地坪

1400 520 1800 2920 4510 2920 1800 520 1400

17790

广胜上寺弥陀殿正立面图
Front Elevation of Amitābha Hall, Upper Guangsheng Monastery

13.140 正吻最高点

0 1 3m

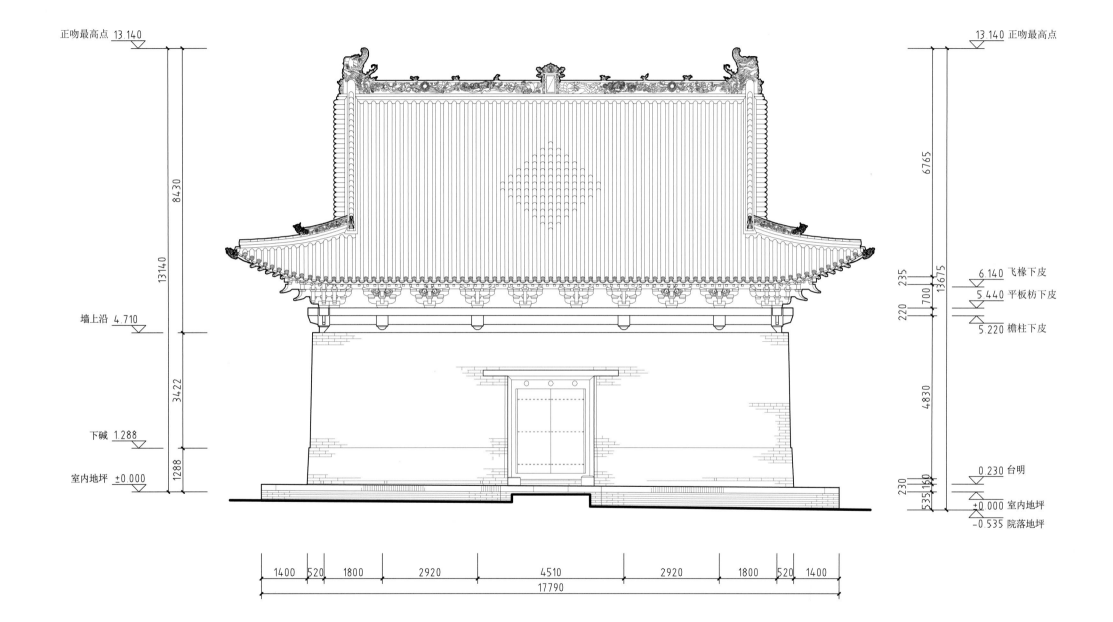

正吻最高点 13.140

墙上沿 4.710

下碱 1.288

室内地坪 ±0.000

8430

13140

3422

1288

13.140 正吻最高点

6765

13675

235

220

700

6.140 飞椽下皮
5.440 平板枋下皮
5.220 檐柱下皮

4830

0.230 台明

230
535 150

+0.000 室内地坪
-0.535 院落地坪

1400 520 1800 2920 4510 2920 1800 520 1400

17790

广胜上寺弥陀殿背立面图
Back Elevation of Amitābha Hall, Upper Guangsheng Monastery

0   1   3m

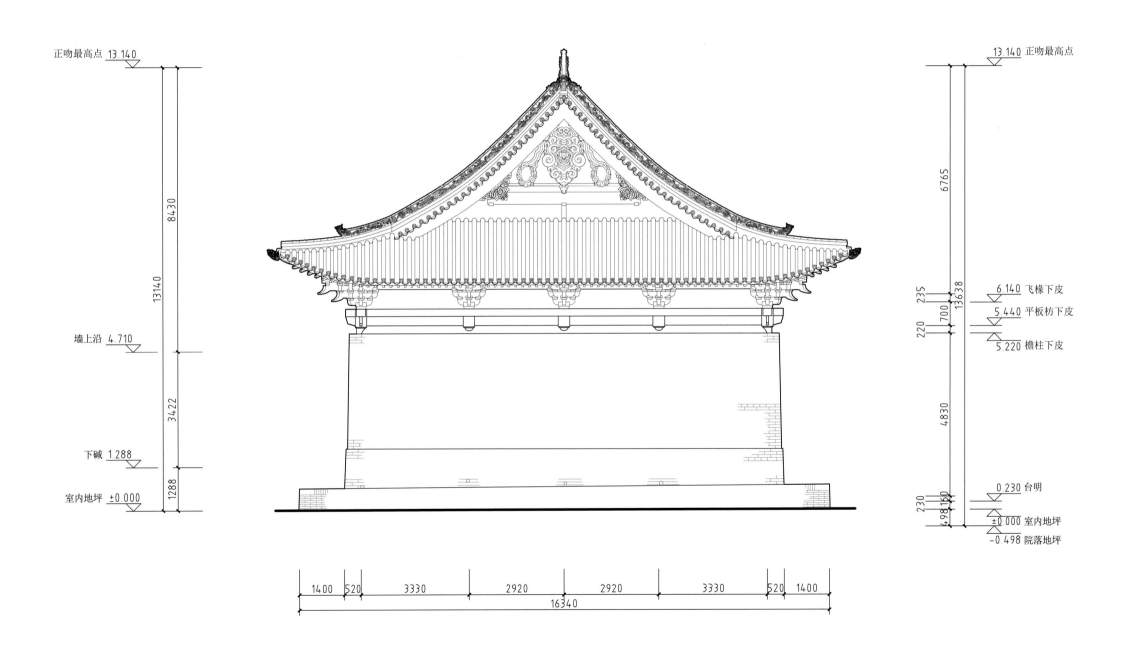

正吻最高点 13.140

墙上沿 4.710

下碱 1.288

室内地坪 ±0.000

8430

13140

3422

1288

正吻最高点 13.140

6765

13638

235

6.140 飞椽下皮

700

5.440 平板枋下皮

220

5.220 檐柱下皮

4830

0.230 台明

230

498 150

±0.000 室内地坪

-0.498 院落地坪

1400　520　3330　2920　2920　3330　520　1400

16340

广胜上寺弥陀殿侧立面图
Side Elevation of Amitābha Hall, Upper Guangsheng Monastery

0　1　3m

347 860　2629　1413　742　2170　2170　742　1413　2629　860 347

正吻最高点 13.140 — 435
宝顶最高点 12.705 — 1615
脊檩上皮 11.090 — 1970
上金檩上皮 9.120 — 710
斜梁上皮 8.410 — 735
下金檩上皮 7.675 — 1042
正心檩上皮 6.632 — 312 — 13140
飞椽下皮 6.320 — 1100
檐柱上皮 5.220 — 510
墙上沿 4.710 — 4480
台明 0.230 — 230
室内地坪 ±0.000

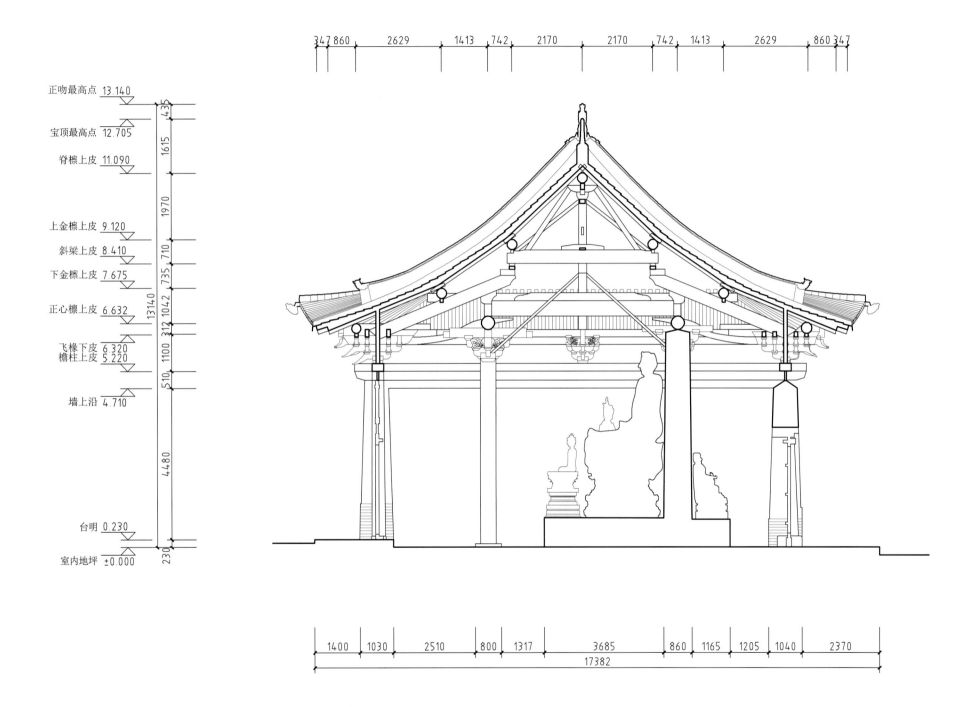

1400　1030　2510　800　1317　3685　860　1165　1205　1040　2370
17382

广胜上寺弥陀殿横剖面图
Cross Cross Section of Amitābha Hall, Upper Guangsheng Monastery

0　1　3m

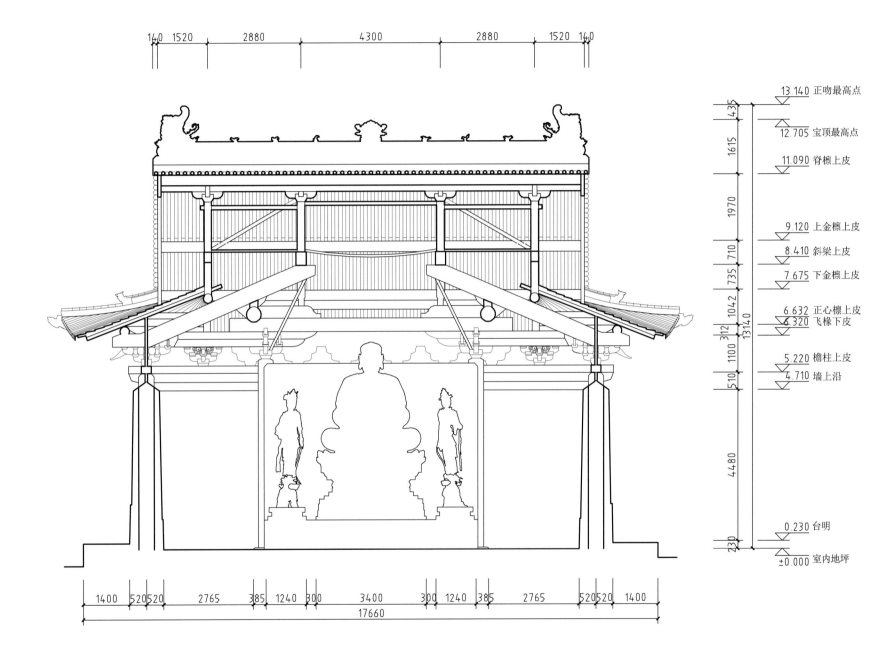

13.140 正吻最高点

12.705 宝顶最高点

11.090 脊槫上皮

9.120 上金槫上皮

8.410 斜梁上皮

7.675 下金槫上皮

6.632 正心槫上皮
6.320 飞椽下皮

5.220 檐柱上皮

4.710 墙上沿

0.230 台明

±0.000 室内地坪

广胜上寺弥陀殿纵剖面图

Longitudinal Longitudinal Section of Amitābha Hall, Upper Guangsheng Monastery

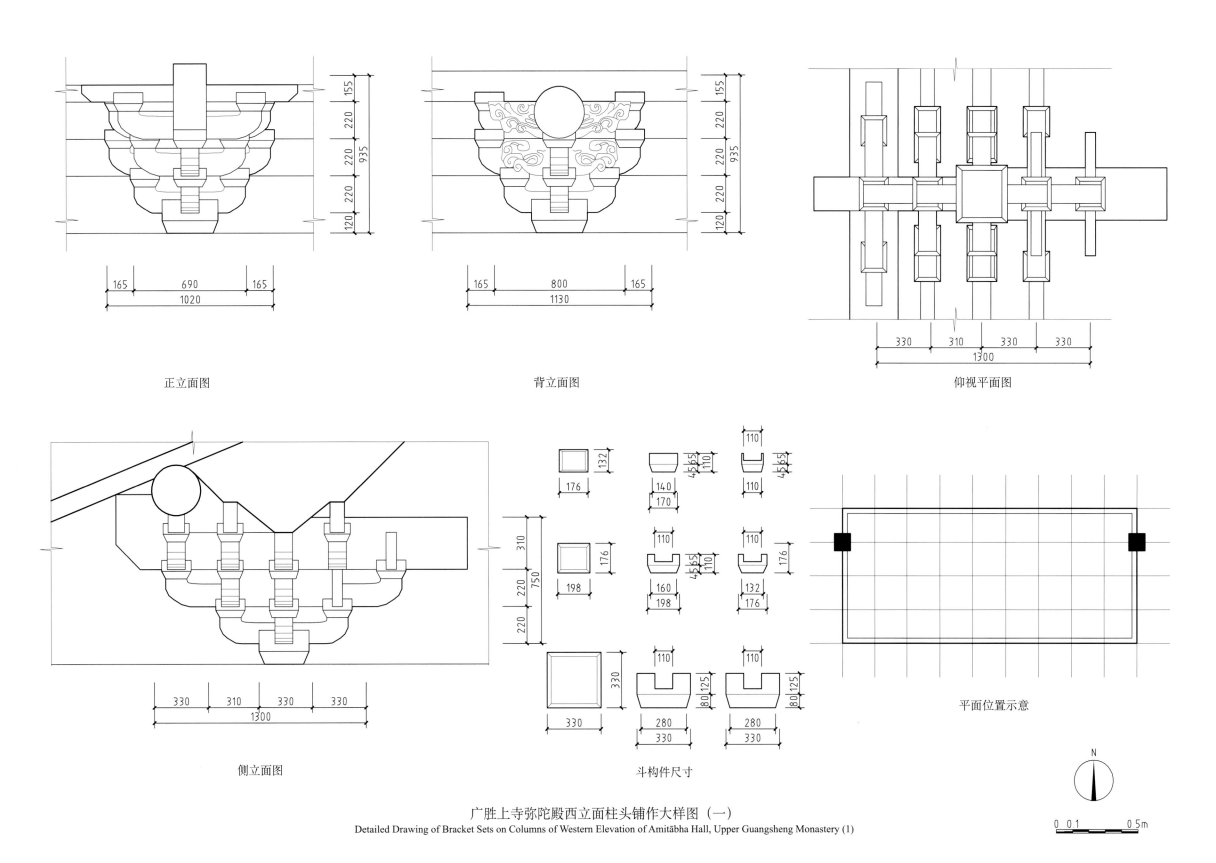

正立面图

背立面图

仰视平面图

侧立面图

斗构件尺寸

平面位置示意

广胜上寺弥陀殿西立面柱头铺作大样图（一）

Detailed Drawing of Bracket Sets on Columns of Western Elevation of Amitābha Hall, Upper Guangsheng Monastery (1)

0 0.1   0.5m

155
220
220
935
120 220 220

165 690 165
1020

正立面图

250
220
1030
220 220
120

165 800 165
1130

背立面图

330 310 330 330
1300

仰视平面图

310
750
220 220

330 330 310 330
1300

侧立面图

132
176

176
198

330
330

斗构件尺寸

平面位置示意

N

广胜上寺弥陀殿西立面柱头铺作大样图（二）
Detailed Drawing of Bracket Sets on Columns of Western Elevation of Amitābha Hall, Upper Guangsheng Monastery (2)

0 0.1   0.5m

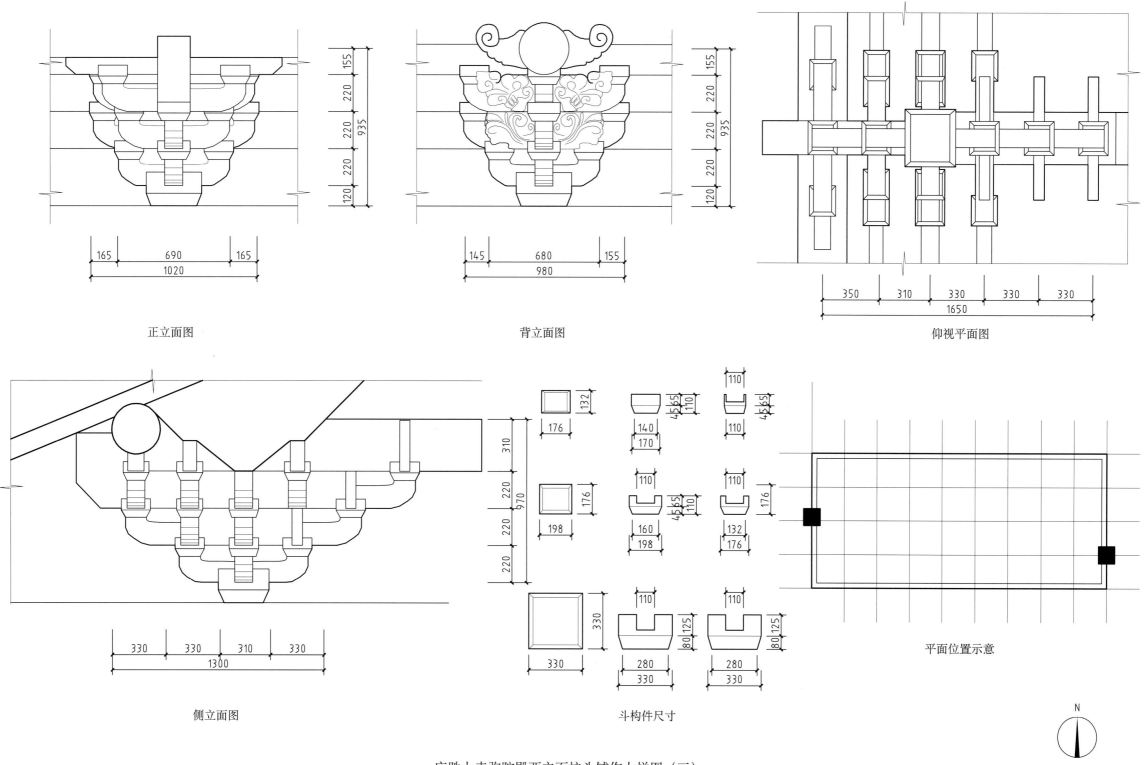

正立面图

背立面图

仰视平面图

侧立面图

斗构件尺寸

平面位置示意

N

广胜上寺弥陀殿西立面柱头铺作大样图（三）
Detailed Drawing of Bracket Sets on Columns of Western Elevation of Amitābha Hall, Upper Guangsheng Monastery (3)

0 0.1 0.5m

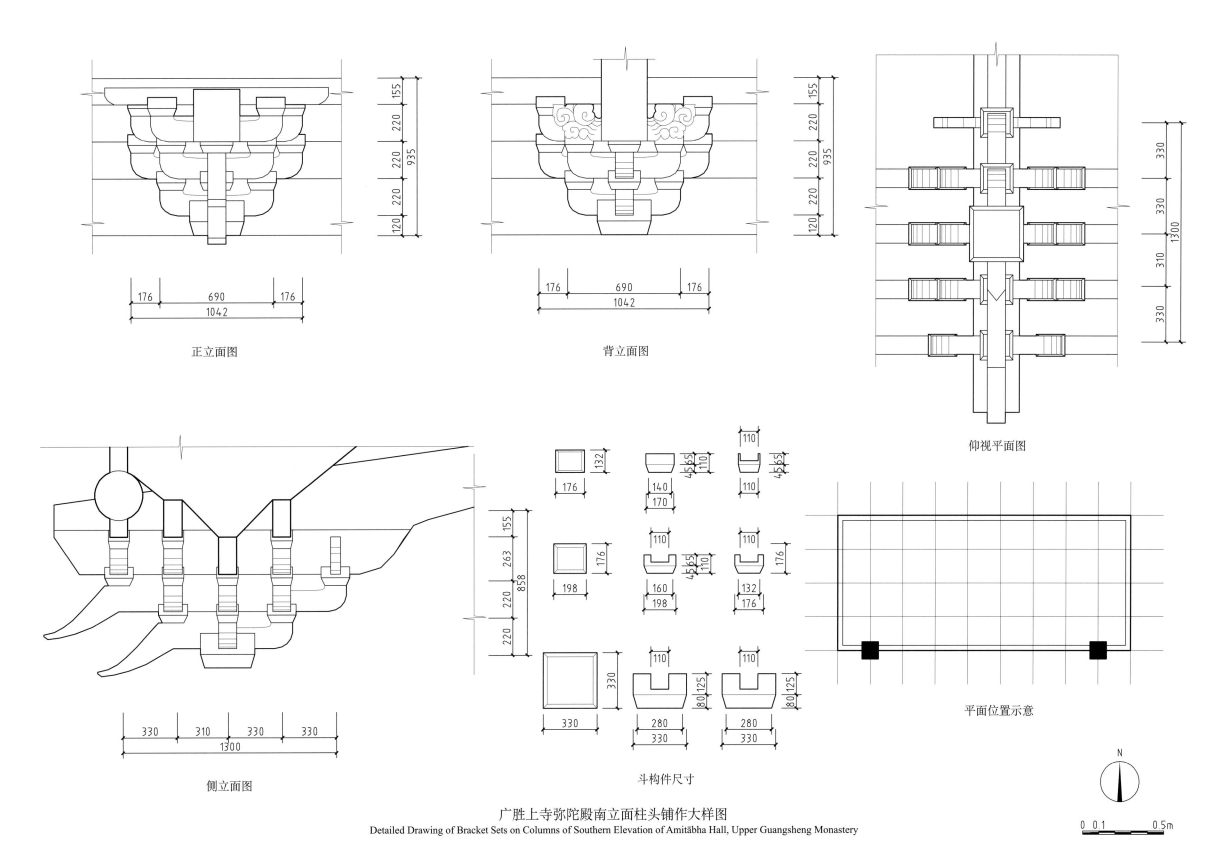

正立面图

背立面图

仰视平面图

侧立面图

斗构件尺寸

平面位置示意

广胜上寺弥陀殿南立面柱头铺作大样图

Detailed Drawing of Bracket Sets on Columns of Southern Elevation of Amitābha Hall, Upper Guangsheng Monastery

0 0.1   0.5 m

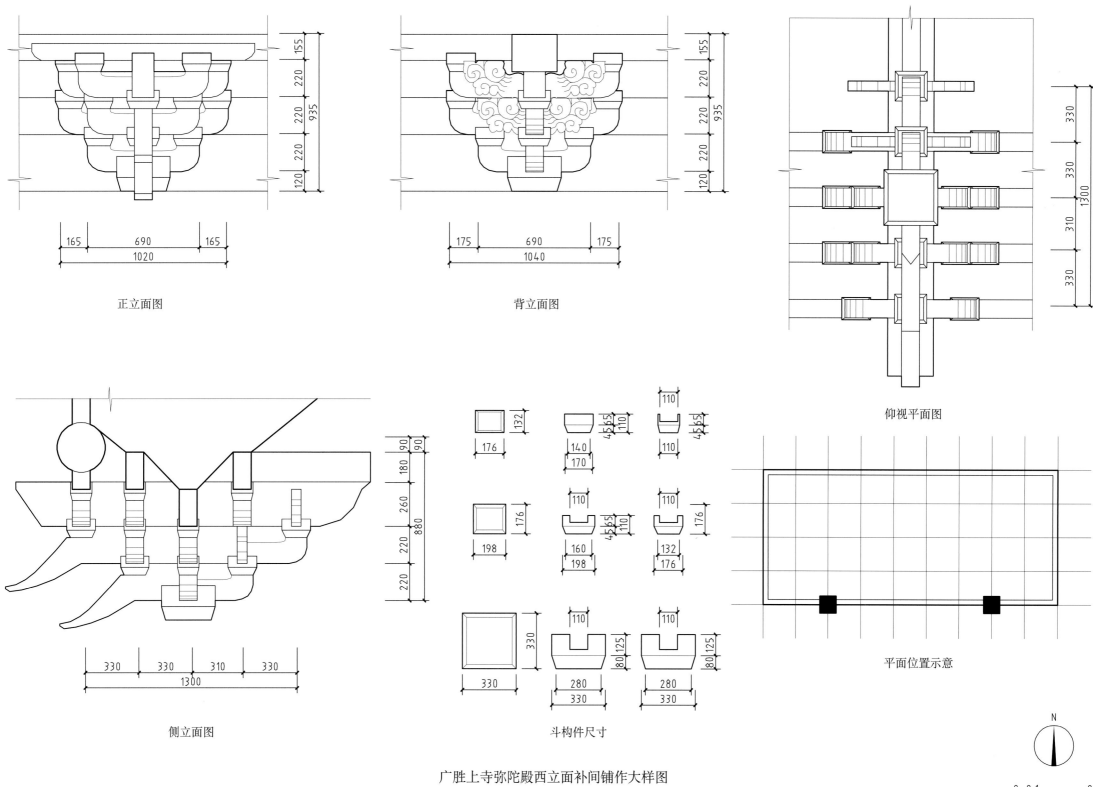

正立面图

背立面图

仰视平面图

侧立面图

斗构件尺寸

平面位置示意

广胜上寺弥陀殿西立面补间铺作大样图

Detailed Drawing of Bracket Sets between Columns of Western Elevation of Amitābha Hall, Upper Guangsheng Monastery

0 0.1   0.5m

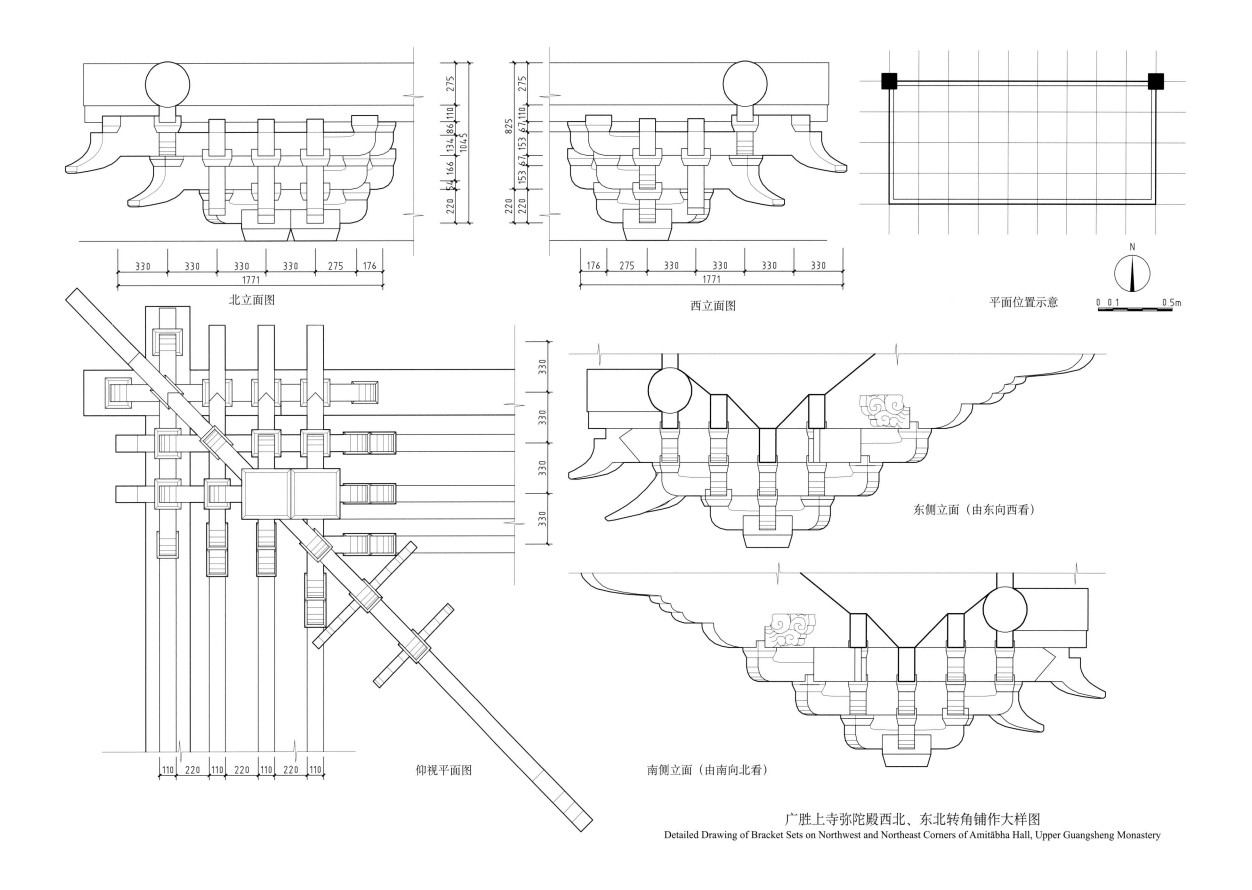

北立面图

西立面图

平面位置示意

N

0 0.1 0.5m

仰视平面图

东侧立面（由东向西看）

南侧立面（由南向北看）

广胜上寺弥陀殿西北、东北转角铺作大样图

Detailed Drawing of Bracket Sets on Northwest and Northeast Corners of Amitābha Hall, Upper Guangsheng Monastery

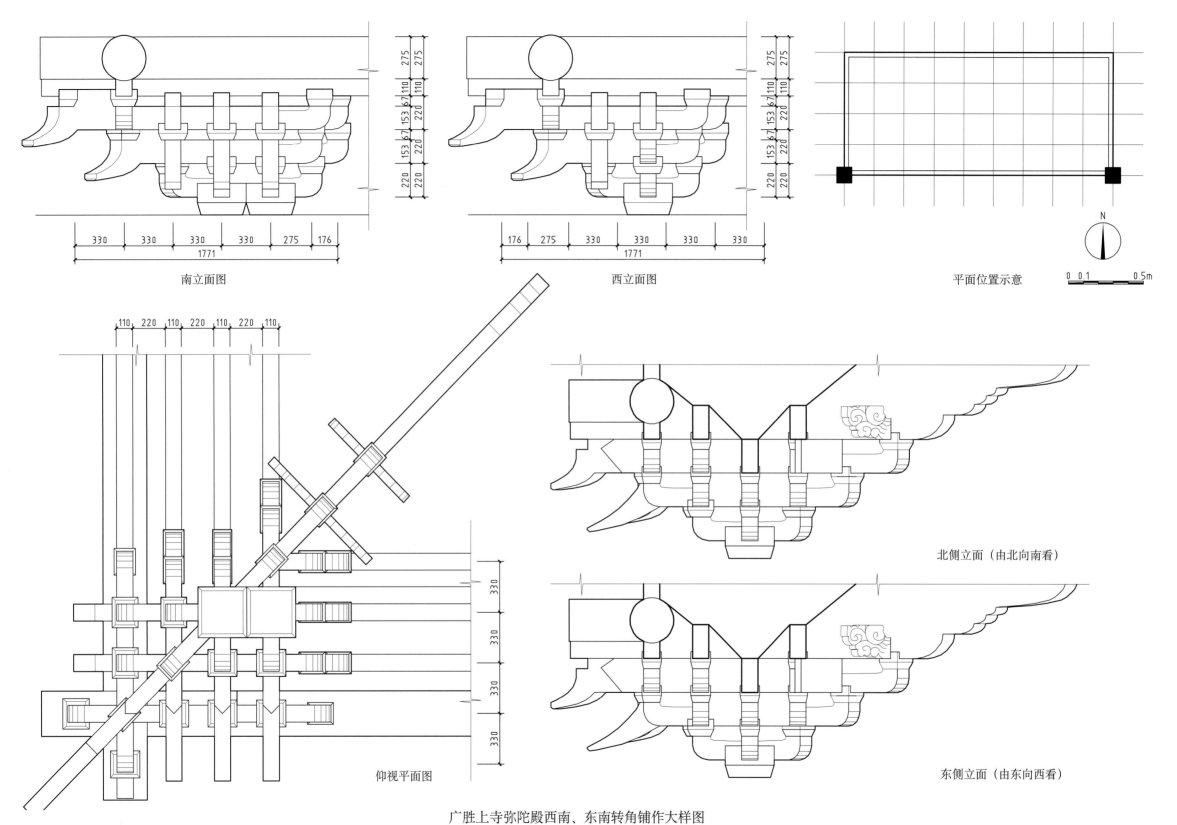

南立面图

西立面图

平面位置示意

仰视平面图

北侧立面（由北向南看）

东侧立面（由东向西看）

广胜上寺弥陀殿西南、东南转角铺作大样图

Detailed Drawing of Bracket Sets on Southwest and Southeast Corners of Amitābha Hall, Upper Guangsheng Monastery

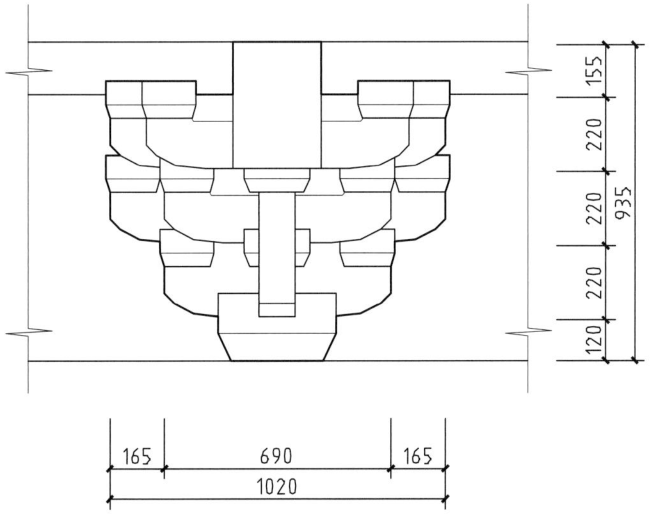

059

正立面图

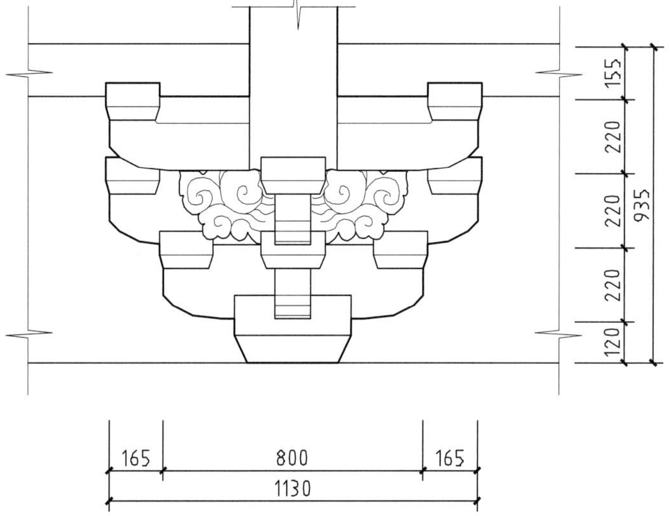

背立面图

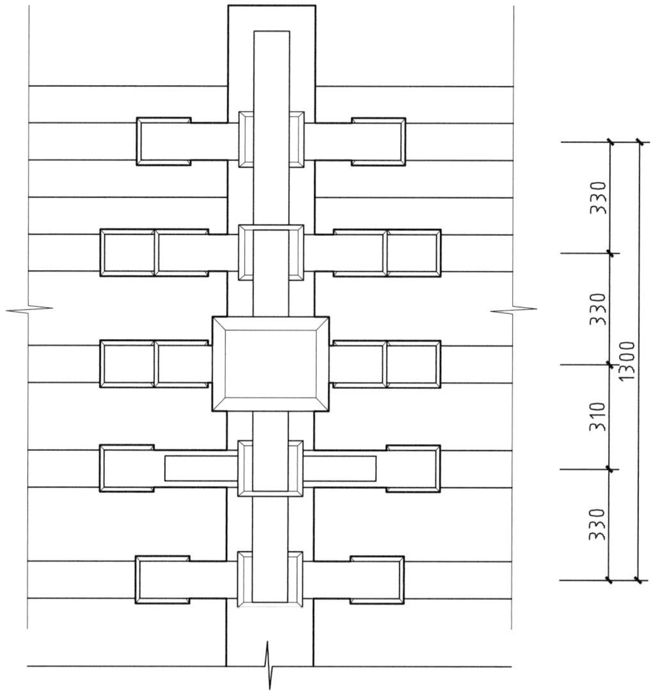

仰视平面图（上方为室外）

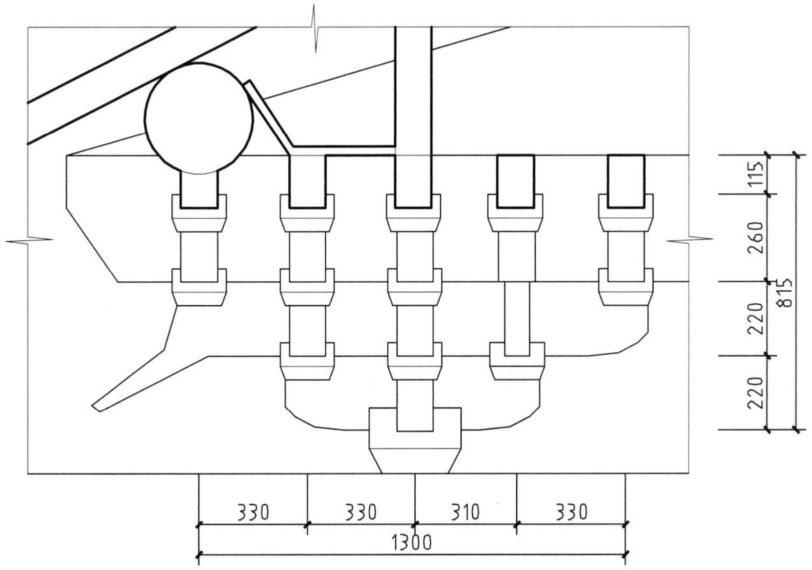

侧立面图

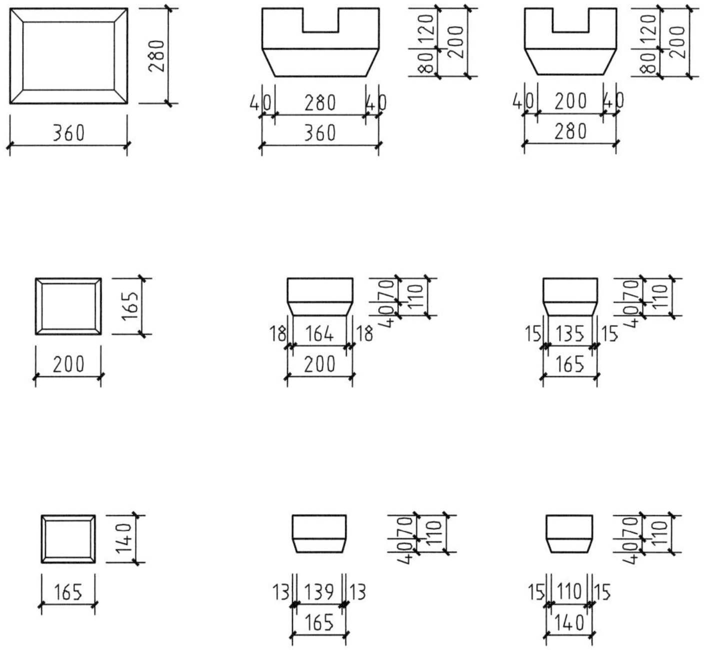

斗构件尺寸

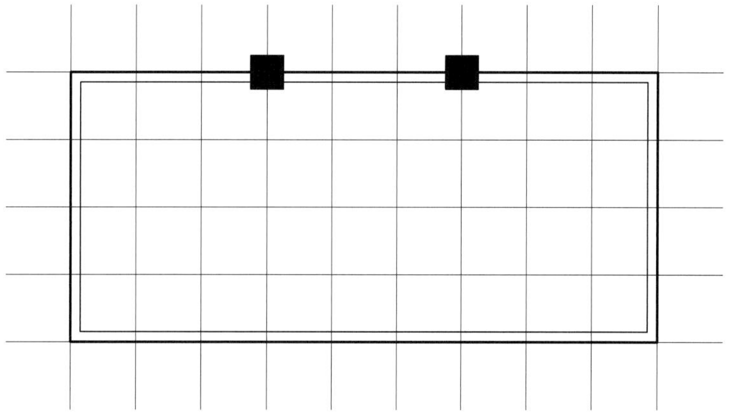

平面位置示意

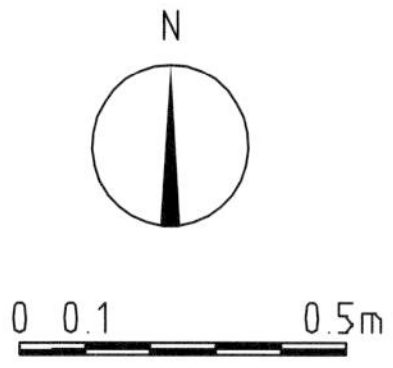

广胜上寺弥陀殿北立面柱头铺作大样图（一）
Detailed Drawing of Bracket Sets on Columns of Northern Elevation of Amitābha Hall, Upper Guangsheng Monastery (1)

0 0.1   0.5m

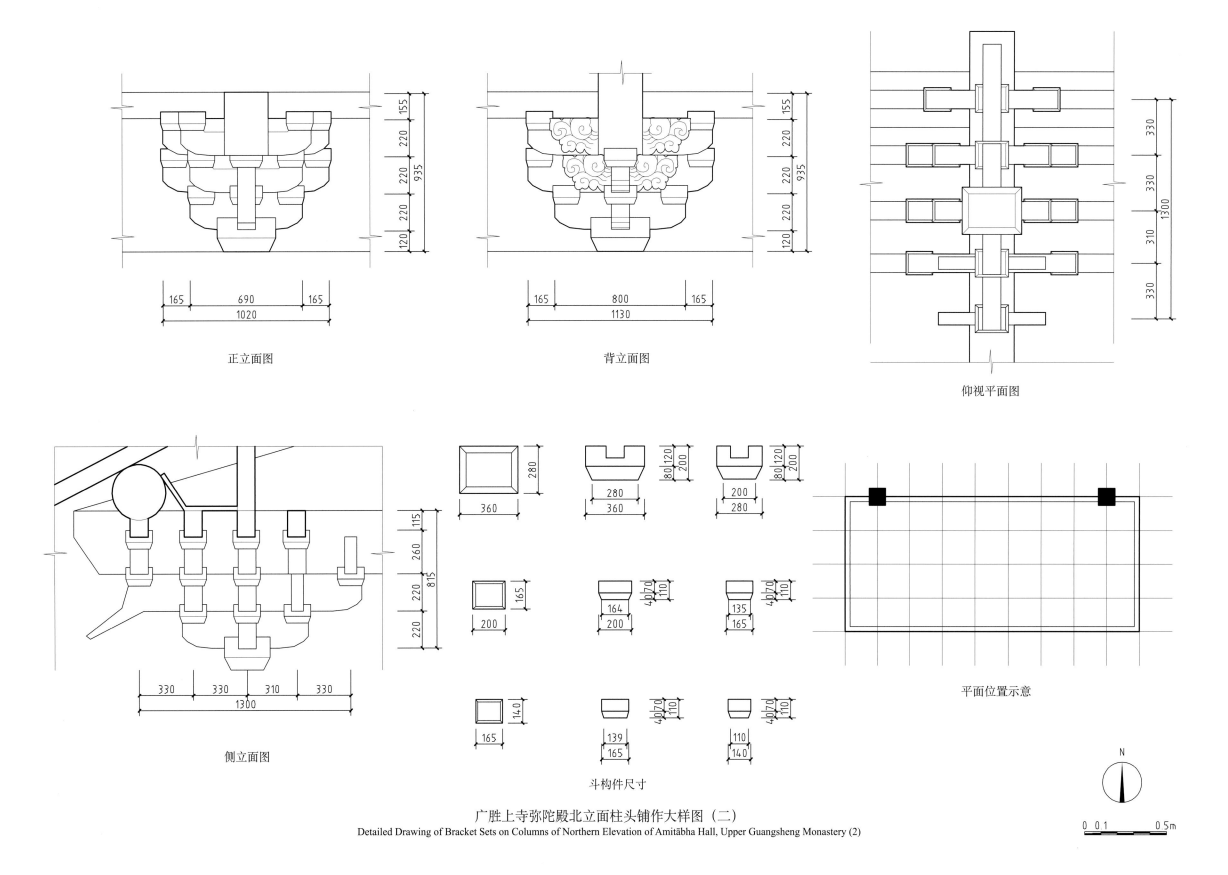

正立面图

背立面图

仰视平面图

侧立面图

斗构件尺寸

平面位置示意

广胜上寺弥陀殿北立面柱头铺作大样图（二）
Detailed Drawing of Bracket Sets on Columns of Northern Elevation of Amitābha Hall, Upper Guangsheng Monastery (2)

N

0 0.1   0.5m

061

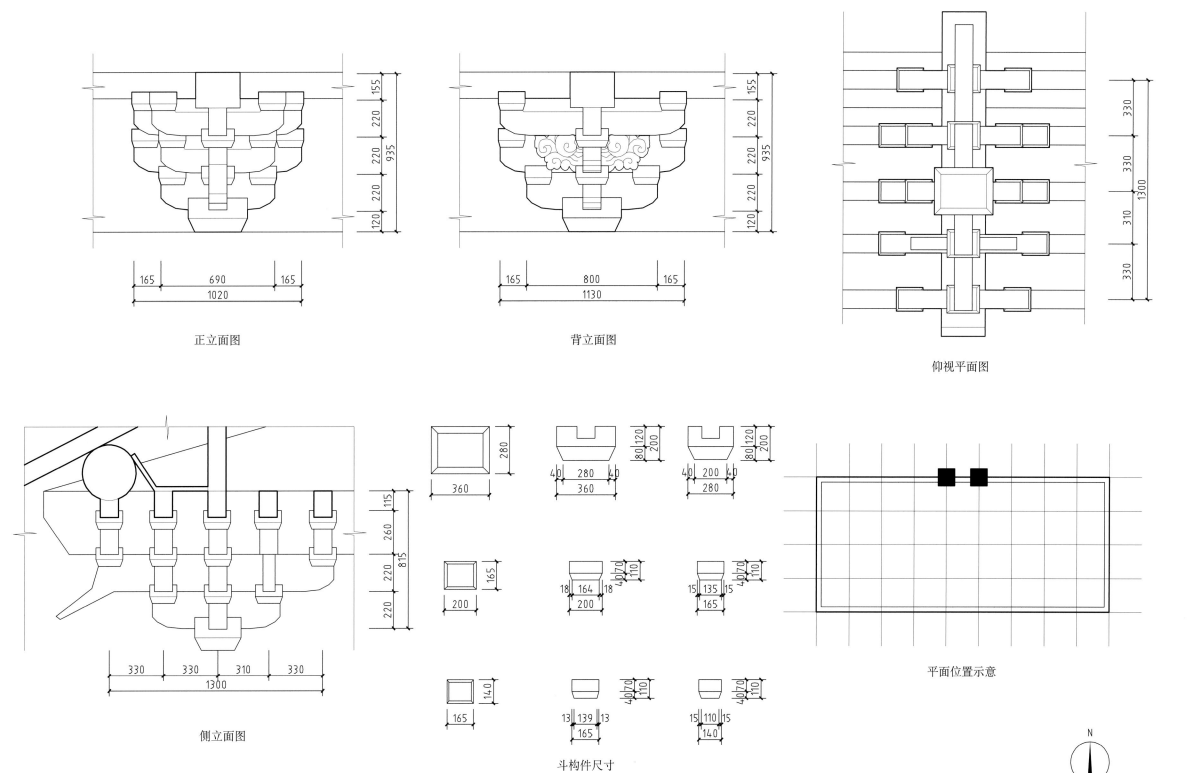

正立面图

背立面图

仰视平面图

侧立面图

斗构件尺寸

平面位置示意

N

广胜上寺弥陀殿北立面补间铺作大样图（一）

Detailed Drawing of Bracket Sets between Columns of Northern Elevation of Amitābha Hall, Upper Guangsheng Monastery (1)

0 0.1   0.5m

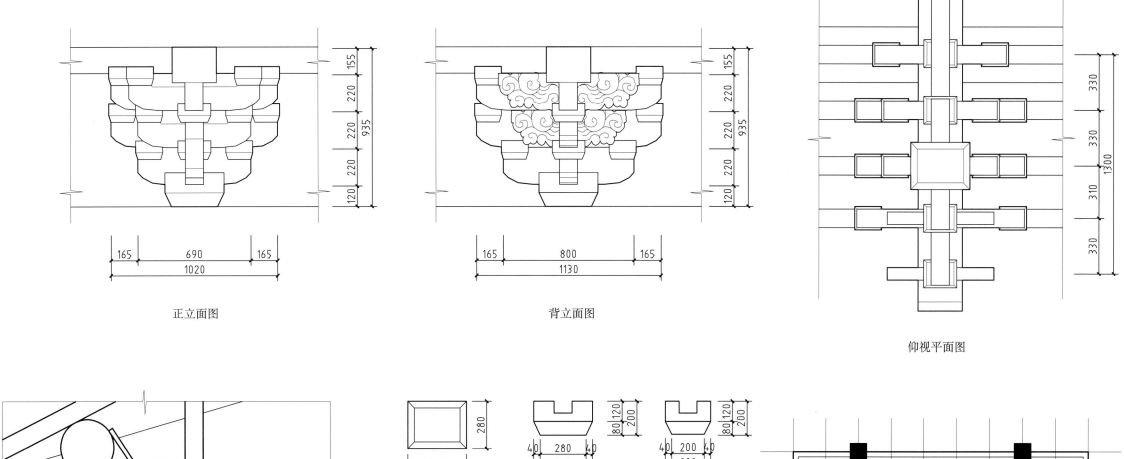

正立面图

背立面图

仰视平面图

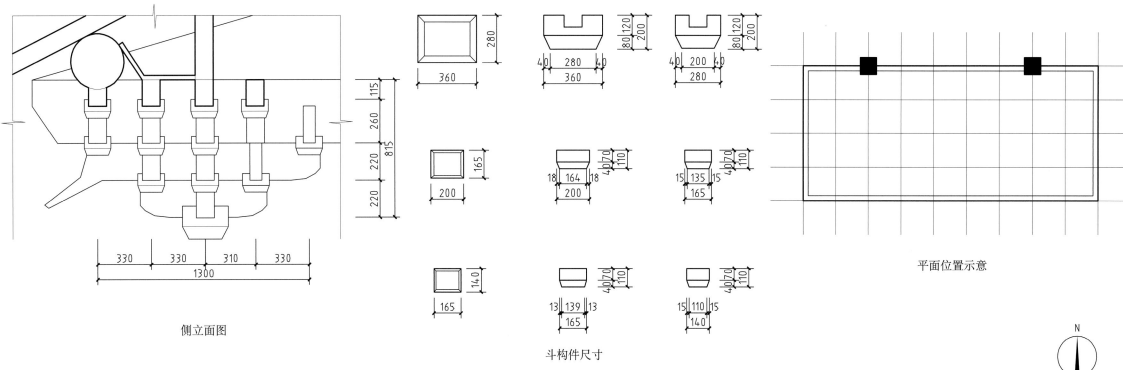

侧立面图

斗构件尺寸

平面位置示意

N

广胜上寺弥陀殿北立面补间铺作大样图（二）
Detailed Drawing of Bracket Sets between Columns of Northern Elevation of Amitābha Hall, Upper Guangsheng Monastery (2)

0 0.1 0.5m

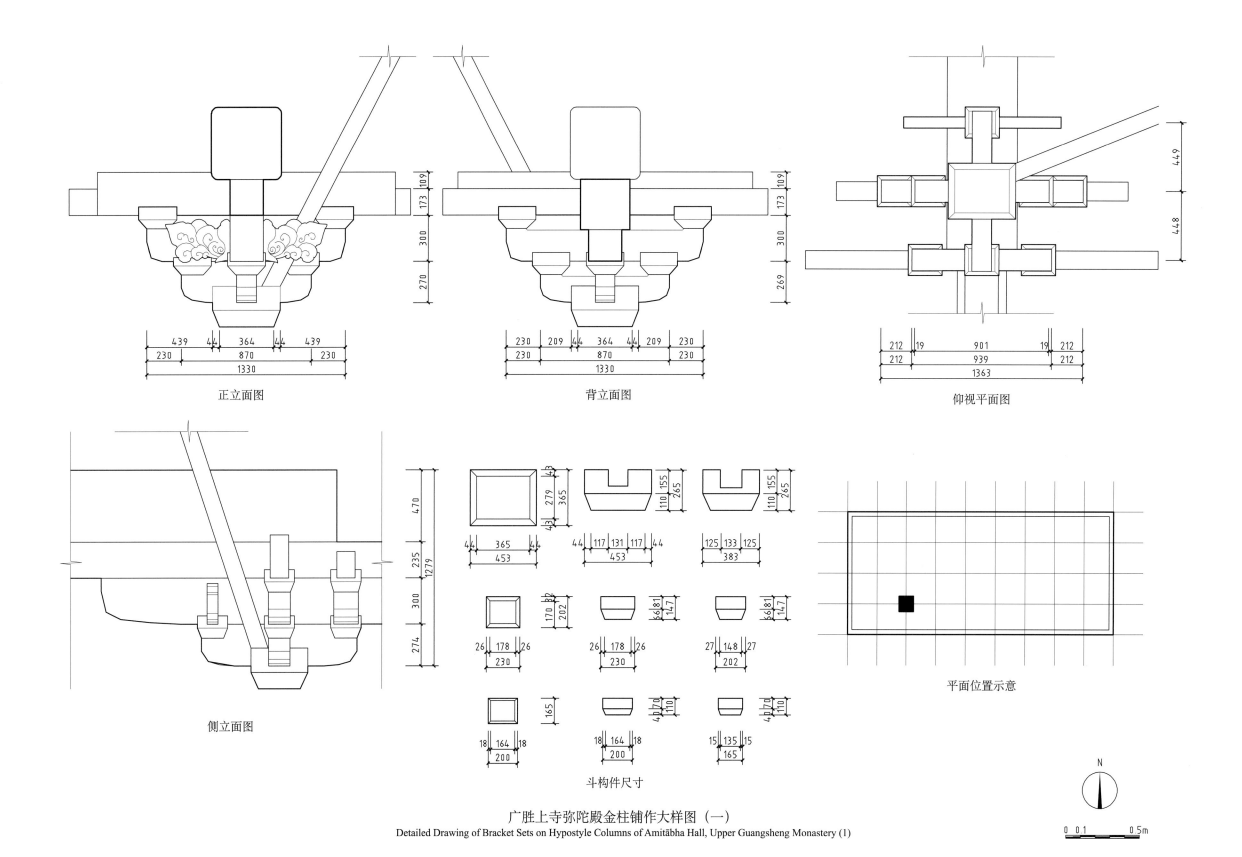

正立面图

背立面图

仰视平面图

侧立面图

斗构件尺寸

平面位置示意

广胜上寺弥陀殿金柱铺作大样图（一）

Detailed Drawing of Bracket Sets on Hypostyle Columns of Amitābha Hall, Upper Guangsheng Monastery (1)

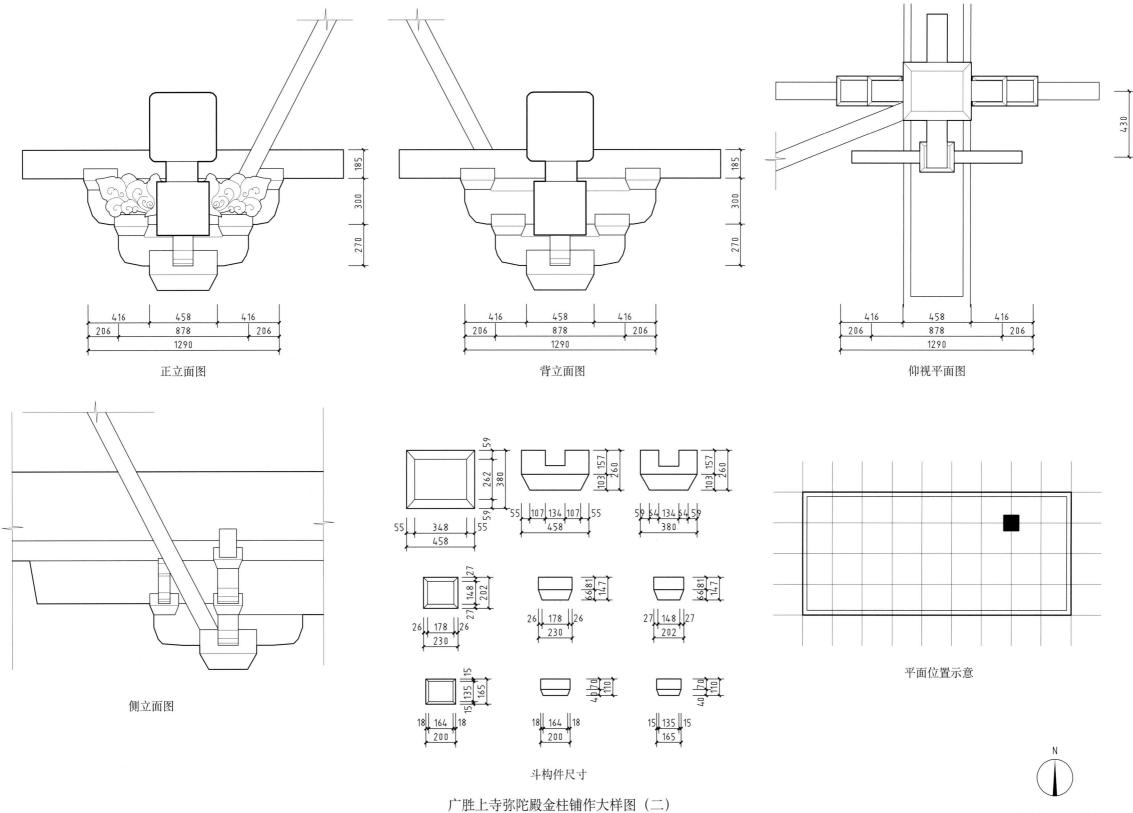

正立面图

背立面图

仰视平面图

侧立面图

斗构件尺寸

平面位置示意

广胜上寺弥陀殿金柱铺作大样图（二）

Detailed Drawing of Bracket Sets on Hypostyle Columns of Amitābha Hall, Upper Guangsheng Monastery (2)

N

0　0.1　　　0.5m

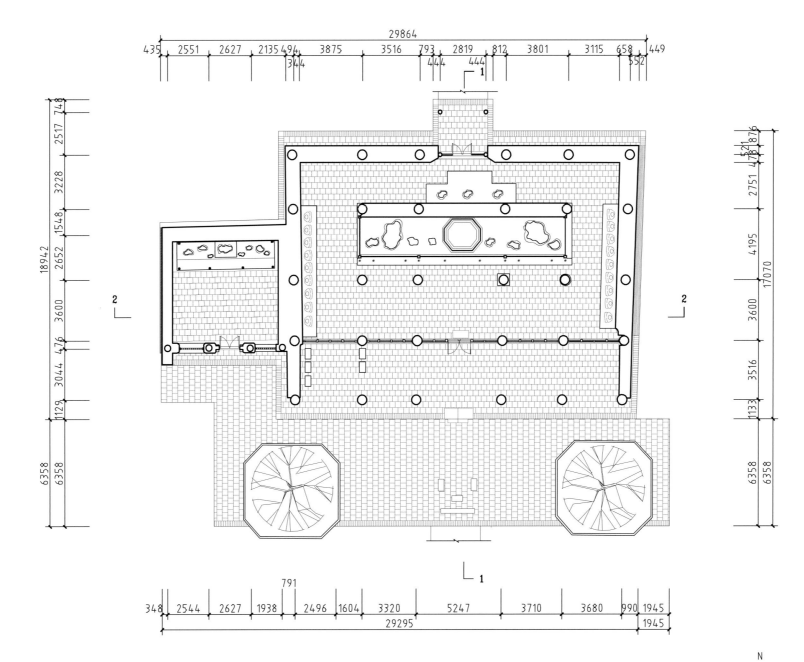

29864

435 2551 2627 2135 494 3875 3516 793 2819 812 3801 3115 658 449
344 444 444 552

2517 748
3228
1548
2652
3600
476
3044
1129
6358 6358

18942

521 876
478
2751
4195
3600
3516
1133
6358 6358

17070

791

348 2544 2627 1938 2496 1604 3320 5247 3710 3680 990 1945
1945

29295

广胜上寺大雄宝殿及西垛殿平面图
Floor Plan of the Treasure Hall of the Great Hero and the West Ear Building, Upper Guangsheng Monastery

N

0 1 5m

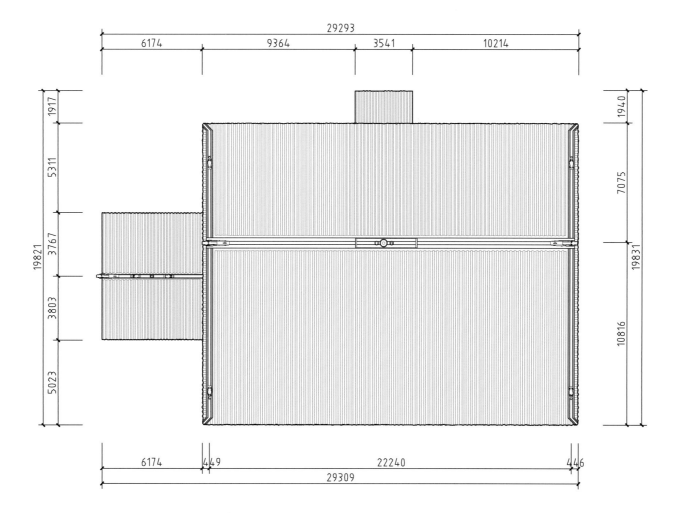

广胜上寺大雄宝殿及西垛殿屋顶平面图
Top View of Roofs of the Treasure Hall of the Great Hero and the West Ear Building, Upper Guangsheng Monastery

0 1 5m

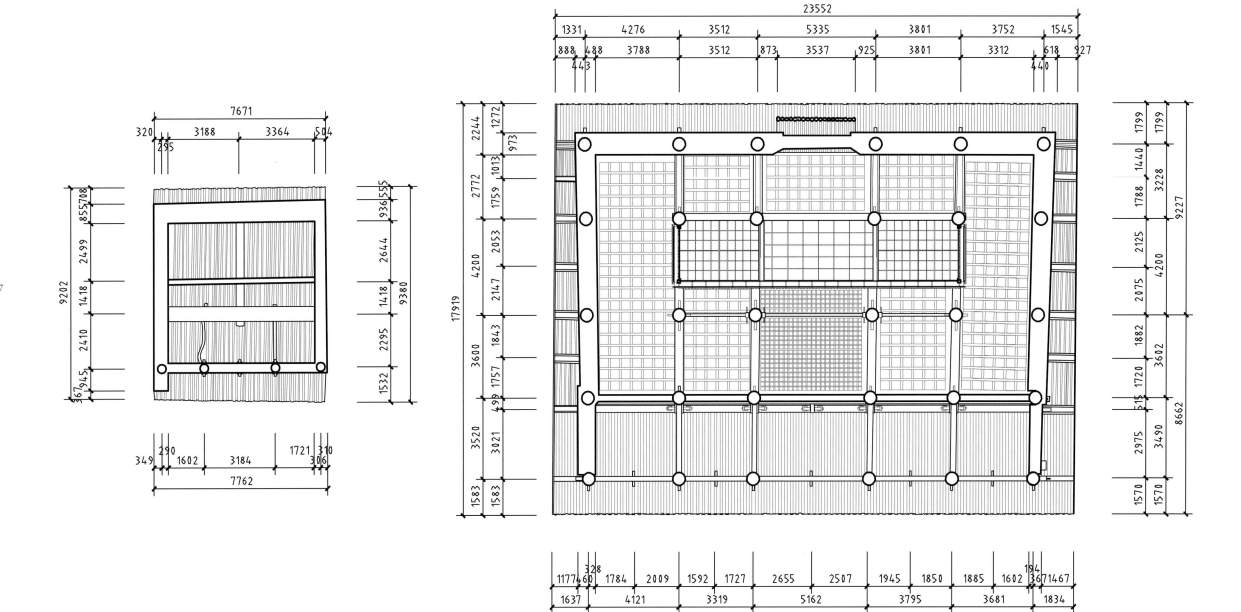

广胜上寺大雄宝殿及西垛殿梁架仰视平面图
Bottom View of Trusses of the Treasure Hall of the Great Hero and the West Ear Building, Upper Guangsheng Monastery

0 1 5m

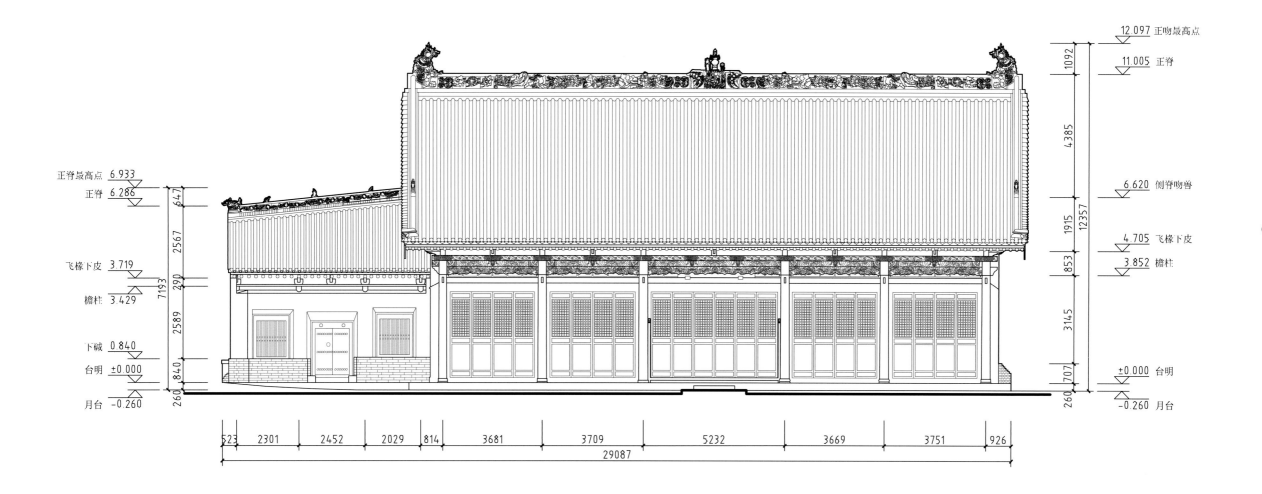

正脊最高点 6.933
正脊 6.286
飞椽卜皮 3.719
檐柱 3.429
卜碱 0.840
台明 ±0.000
月台 -0.260

647
2567
290
7193
2589
840
260

12.097 正吻最高点
11.005 正脊
6.620 侧脊吻兽
4.705 飞椽卜皮
3.852 檐柱
±0.000 台明
-0.260 月台

1092
4.385
1915
12357
853
3145
707
260

523　2301　2452　2029　814　3681　3709　5232　3669　3751　926
29087

广胜上寺大雄宝殿及西垛殿正立面图
Front Elevation of the Treasure Hall of the Great Hero and the West Ear Building, Upper Guangsheng Monastery

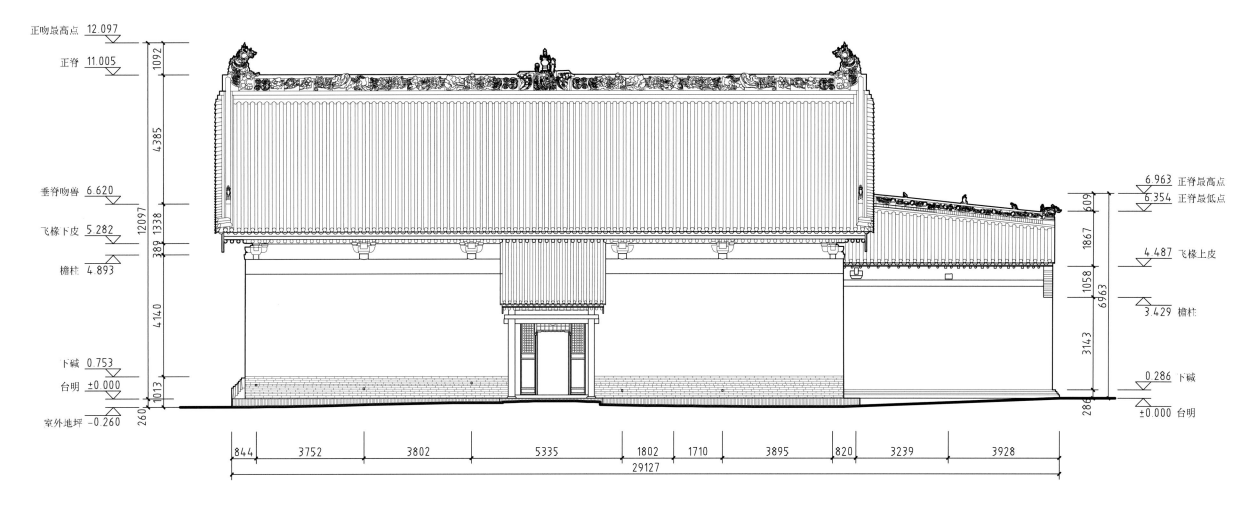

正吻最高点 12.097
正脊 11.005
1092
4385
垂脊吻兽 6.620
飞椽下皮 5.282
12097 1338
389
檐柱 4.893
4140
下碱 0.753
台明 ±0.000
1013
室外地坪 -0.260
260

6.963 正脊最高点
6.354 正脊最低点
609
1867
4.487 飞椽上皮
1058
3.429 檐柱
6963
3143
0.286 下碱
286
±0.000 台明

844　3752　3802　5335　1802　1710　3895　820　3239　3928
29127

广胜上寺大雄宝殿及西垛殿背立面图
Back Elevation of the Treasure Hall of the Great Hero and the West Ear Building, Upper Guangsheng Monastery

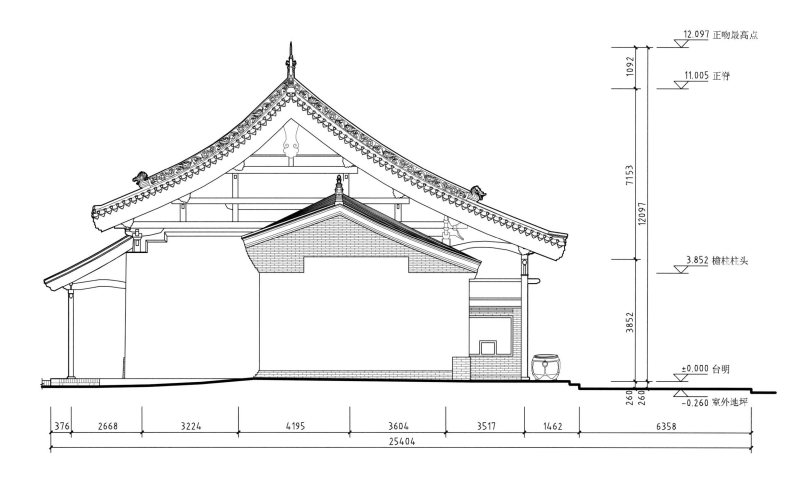

12.097 正吻最高点

11.005 正脊

3.852 檐柱柱头

±0.000 台明

-0.260 室外地坪

1092

7153

12097

3852

260 260

376　2668　3224　4195　3604　3517　1462　6358

25404

广胜上寺大雄宝殿及西垛殿侧立面图
Side Elevation of the Treasure Hall of the Great Hero and the West Ear Building, Upper Guangsheng Monastery

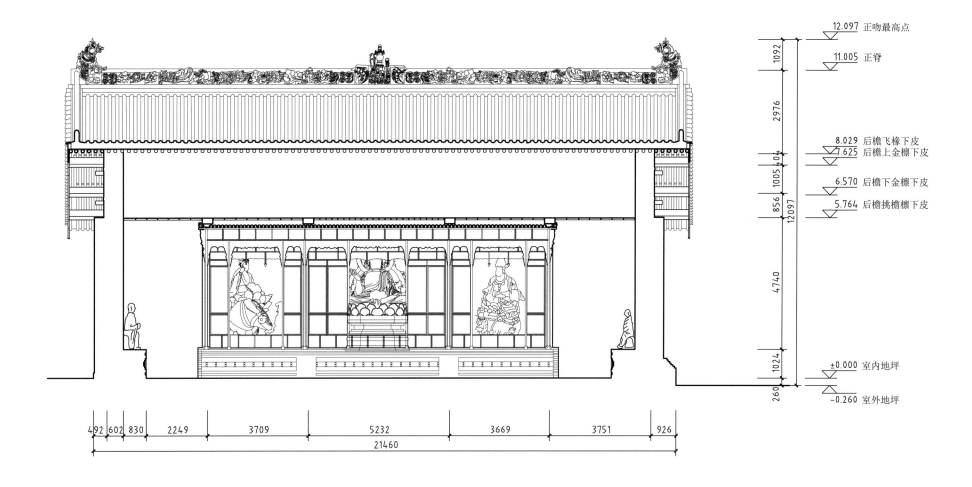

12.097 正吻最高点

11.005 正脊

8.029 后檐飞椽下皮
7.625 后檐上金檩下皮

6.570 后檐下金檩下皮

5.764 后檐挑檐檩下皮

±0.000 室内地坪

-0.260 室外地坪

1092
2976
1005.04
856
4740
1024
260
12097

492 602 830 2249 3709 5232 3669 3751 926
21460

广胜上寺大雄宝殿纵剖面图
Longitudinal Section of the Treasure Hall of the Great Hero, Upper Guangsheng Monastery

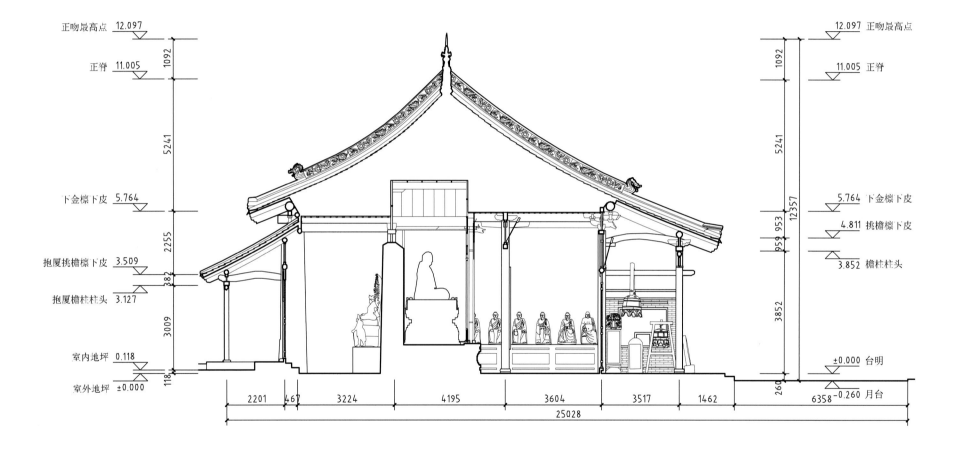

正吻最高点 12.097
正脊 11.005
下金槫下皮 5.764
抱厦挑檐槫下皮 3.509
抱厦檐柱柱头 3.127
室内地坪 0.118
室外地坪 ±0.000

1092
5241
2255
382
3009
118

12.097 正吻最高点
11.005 正脊
5.764 下金槫下皮
4.811 挑檐槫下皮
3.852 檐柱柱头
±0.000 台明
-0.260 月台

1092
5241
953
959
12357
3852
260

2201 467 3224 4195 3604 3517 1462
6358
25028

广胜上寺大雄宝殿横剖面图
Cross Section of the Treasure Hall of the Great Hero, Upper Guangsheng Monastery

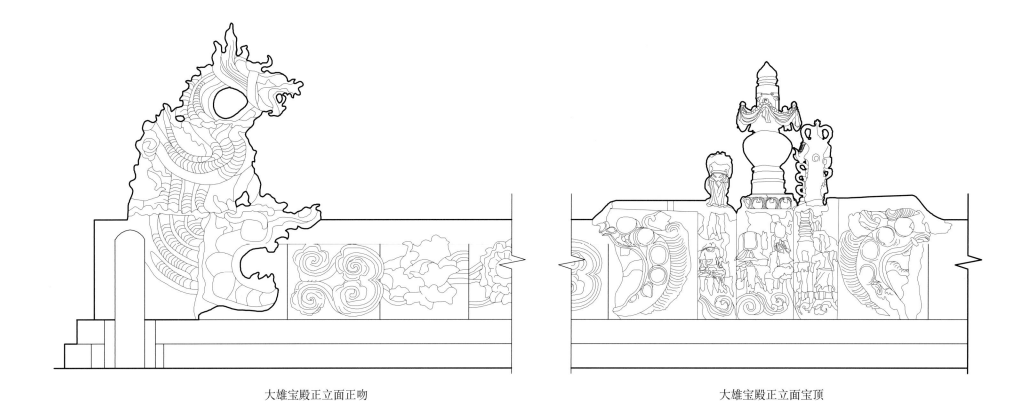

大雄宝殿正立面正吻

大雄宝殿正立面宝顶

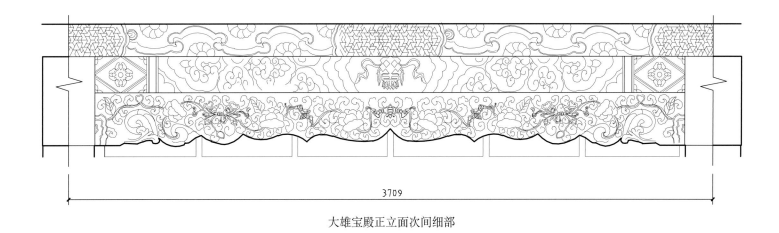

3709

大雄宝殿正立面次间细部

广胜上寺大雄宝殿细部大样图

Detailed Drawing of the Treasure Hall of the Great Hero, Upper Guangsheng Monastery

0　1　3m

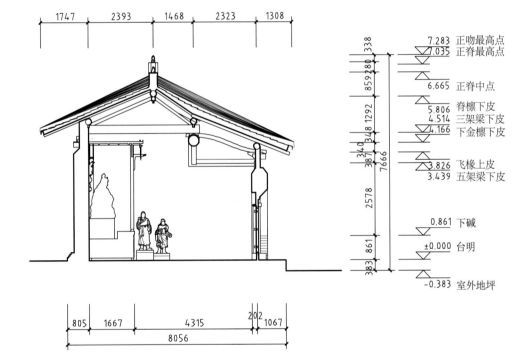

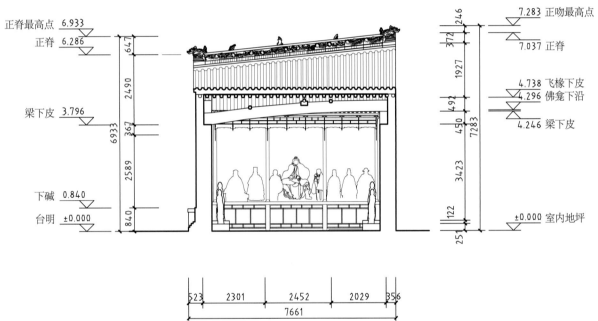

1747　2393　1468　2323　1308

338　7.283　正吻最高点
280　7.035　正脊最高点
859　6.665　正脊中点
1292　5.806　脊檩下皮
348　4.514　三架梁下皮
340　4.166　下金檩下皮
387　3.826　飞椽上皮
2578　3.439　五架梁下皮
7666
861　0.861　下碱
±0.000　台明
383　-0.383　室外地坪

805　1667　4315　202　1067
8056

正脊最高点　6.933
正脊　6.286
647
2490
梁下皮　3.796
6933　367
2589
下碱　0.840
台明　±0.000
840

523　2301　2452　2029　356
7661

246　7.283　正吻最高点
372　7.037　正脊
1927　4.738　飞椽下皮
492　4.296　佛龛下沿
450　4.246　梁下皮
7283
3423
122　±0.000　室内地坪
251

广胜上寺大雄宝殿西垛殿横剖面图
Cross Section of the Treasure Hall of the Great Hero and the West Ear Building,
Upper Guangsheng Monastery

广胜上寺大雄宝殿西垛殿纵剖面图
Longitudinal Section of the Treasure Hall of the Great Hero and the West Ear Building,
Upper Guangsheng Monastery

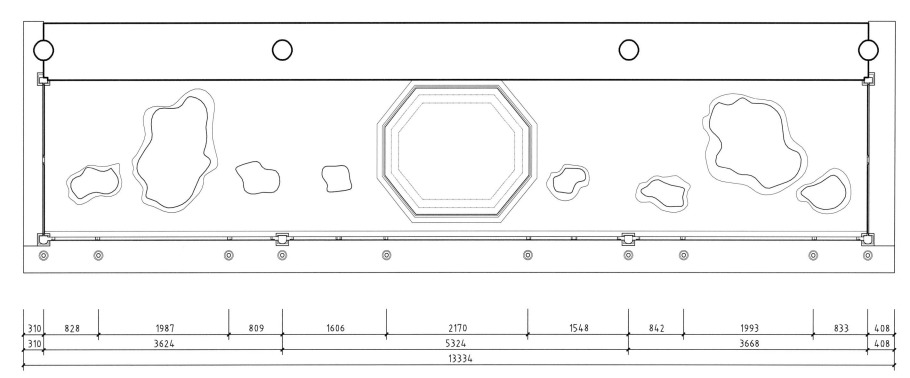

广胜上寺大雄宝殿佛龛平面图
Plan of the Buddha Statue Niche in the Treasure Hall of the Great Hero, Upper Guangsheng Monastery

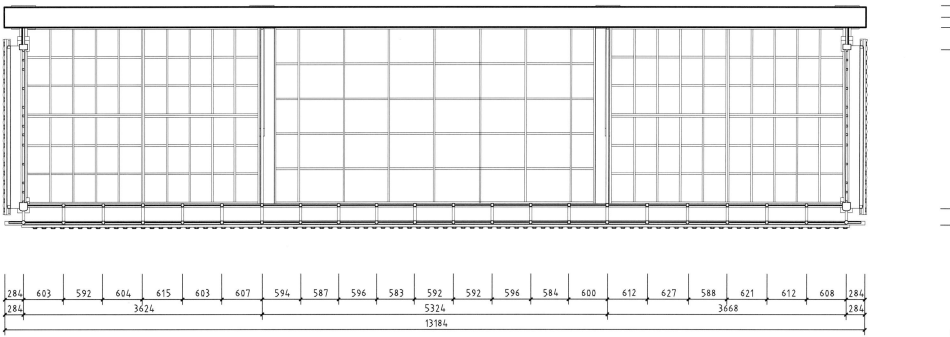

广胜上寺大雄宝殿佛龛仰视平面图
Bottom View of the Buddha Statue Niche in the Treasure Hall of the Great Hero, Upper Guangsheng Monastery

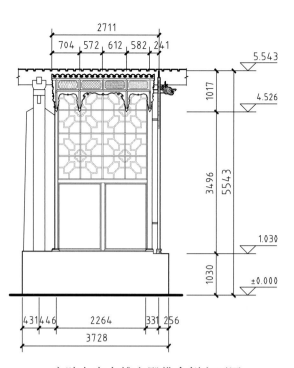

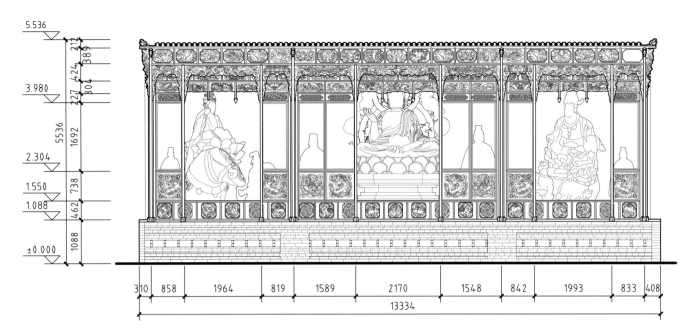

广胜上寺大雄宝殿佛龛正立面图

Front Elevation of the Buddha Statue Niche in the Treasure Hall of the Great Hero, Upper Guangsheng Monastery

广胜上寺大雄宝殿佛龛侧立面图

Side Elevation of the Buddha Statue Niche in the Treasure Hall of the Great Hero, Upper Guangsheng Monastery

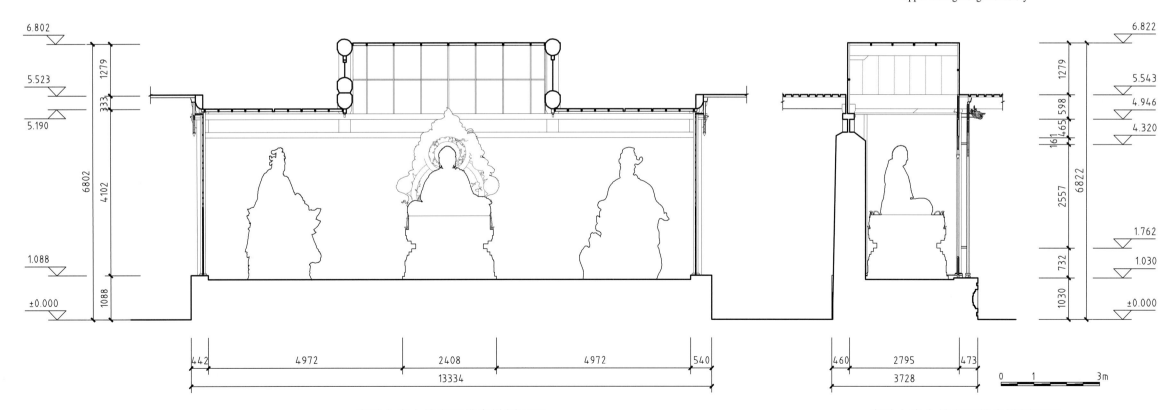

广胜上寺大雄宝殿佛龛纵剖面图

Longitudinal Section of the Buddha Statue Niche in the Treasure Hall of the Great Hero, Upper Guangsheng Monastery

广胜上寺大雄宝殿佛龛横剖面图

Cross Section of the Buddha Statue Niche in the Treasure Hall of the Great Hero, Upper Guangsheng Monastery

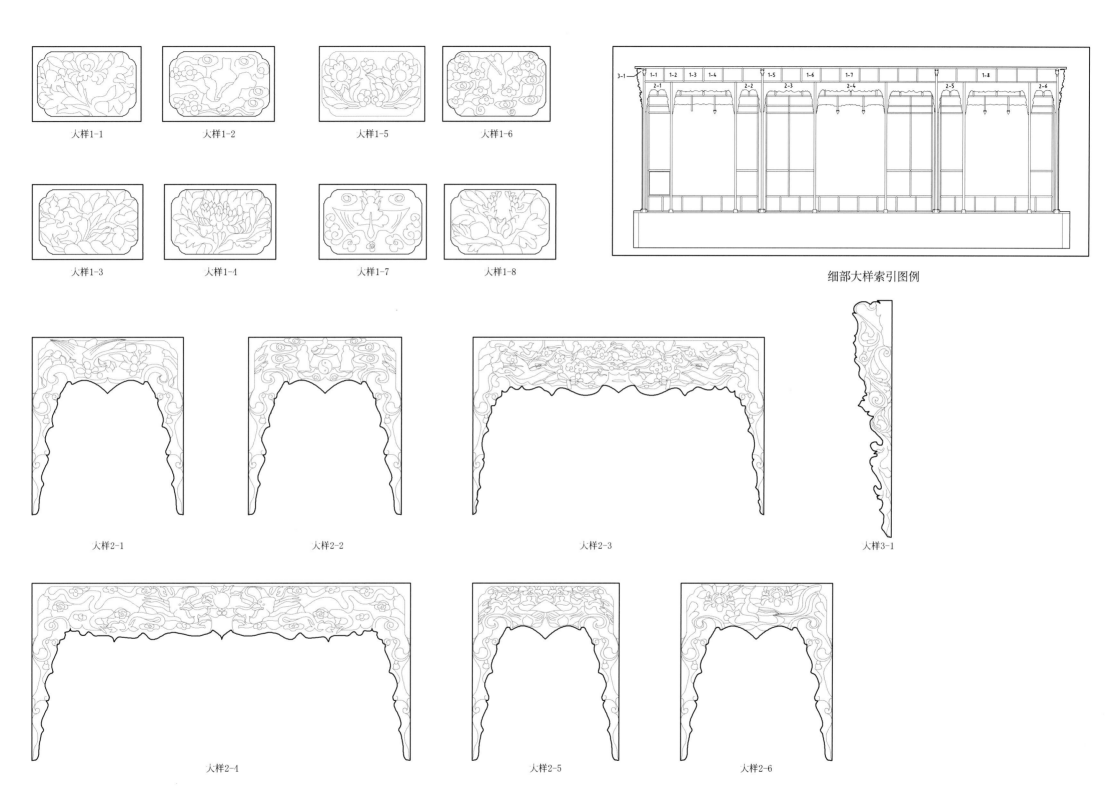

077

大样1-1

大样1-2

大样1-5

大样1-6

大样1-3

大样1-4

大样1-7

大样1-8

细部大样索引图例

大样2-1

大样2-2

大样2-3

大样3-1

大样2-1

大样2-5

大样2-6

广胜上寺大雄宝殿佛龛细部大样图（一）

Detailed Drawing of the Buddha Statue Niche in the Treasure Hall of the Great Hero, Upper Guangsheng Monastery (1)

0　0.2　1m

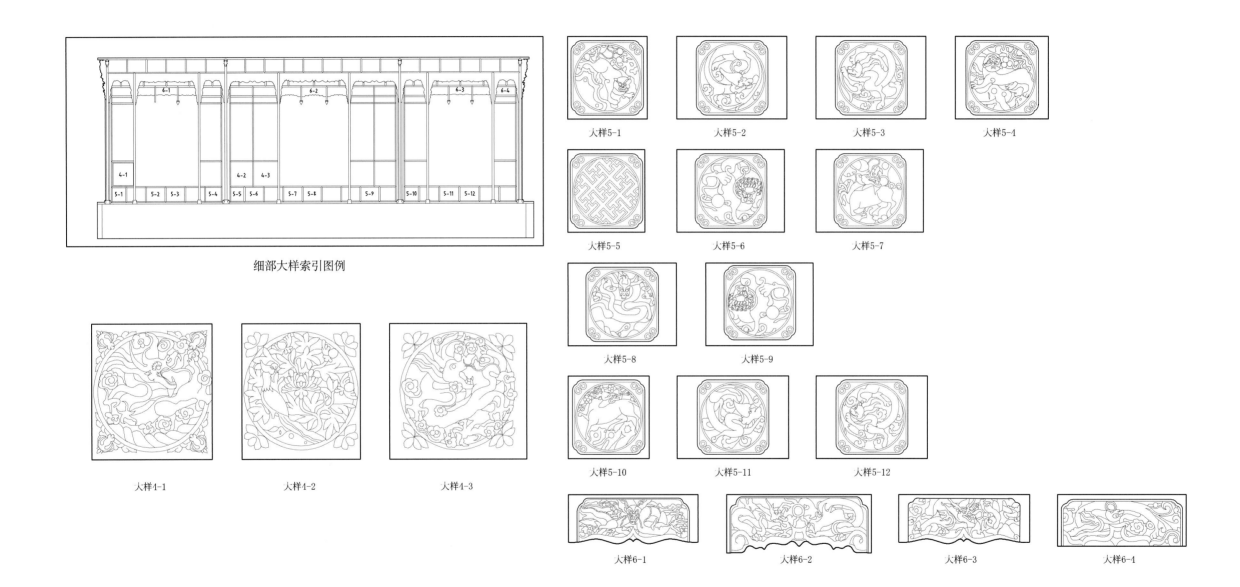

细部大样索引图例

大样4-1　　　　　　大样4-2　　　　　　大样4-3

大样5-1　　　　大样5-2　　　　大样5-3　　　　大样5-4

大样5-5　　　　大样5-6　　　　大样5-7

大样5-8　　　　大样5-9

大样5-10　　　大样5-11　　　大样5-12

大样6-1　　　　大样6-2　　　　大样6-3　　　　大样6-4

广胜上寺大雄宝殿佛龛细部大样图（二）
Detailed Drawing of the Buddha Statue Niche in the Treasure Hall of the Great Hero, Upper Guangsheng Monastery (2)

0　　0.2　　　　　　　　1m

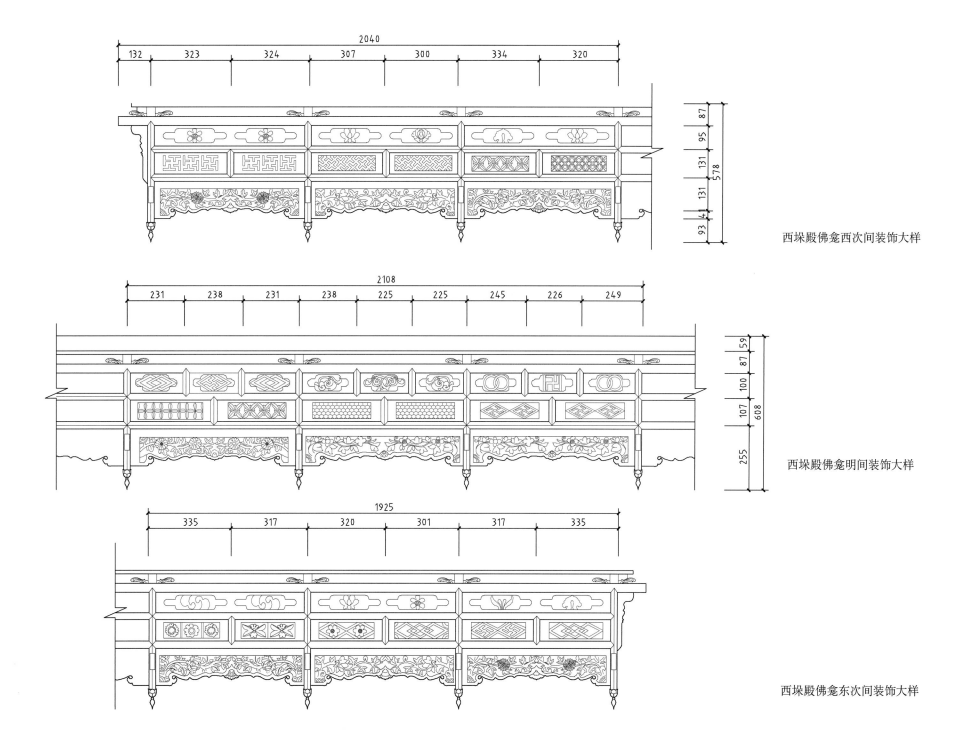

西埝殿佛龛西次间装饰大样

西埝殿佛龛明间装饰大样

西埝殿佛龛东次间装饰大样

广胜上寺大雄宝殿佛龛细部大样图（三）
Detailed Drawing of the Buddha Statue Niche in the Treasure Hall of the Great Hero, Upper Guangsheng Monastery (3)

0   0.1   0.5m

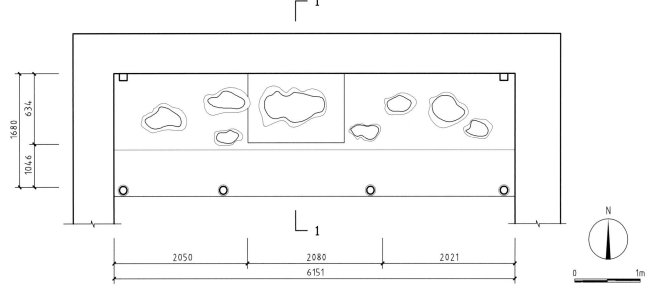

广胜上寺大雄宝殿西垛殿佛龛平面图

Plan of the Buddha Statue Niche in the West Ear Building of the Treasure Hall of the Great Hero, Upper Guangsheng Monastery

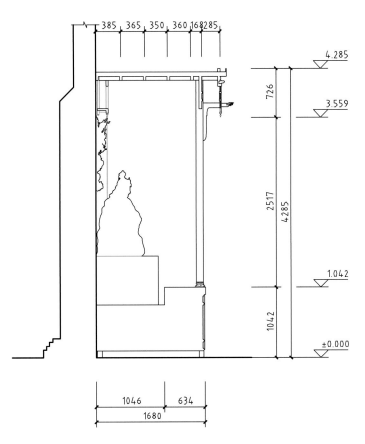

广胜上寺大雄宝殿西垛殿佛龛横剖面图

Cross Section of the Buddha Statue Niche in the West Ear Building of the Treasure Hall of the Great Hero, Upper Guangsheng Monastery

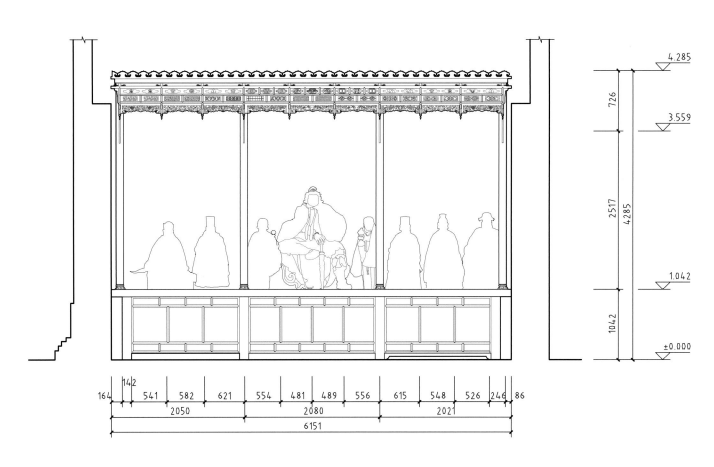

广胜上寺大雄宝殿西垛殿佛龛正立面图

Front Elevation of the Buddha Statue Niche in the West Ear Building of the Treasure Hall of the Great Hero, Upper Guangsheng Monastery

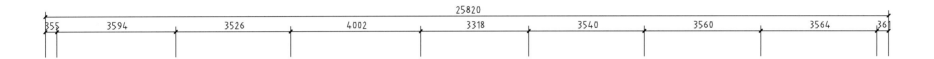

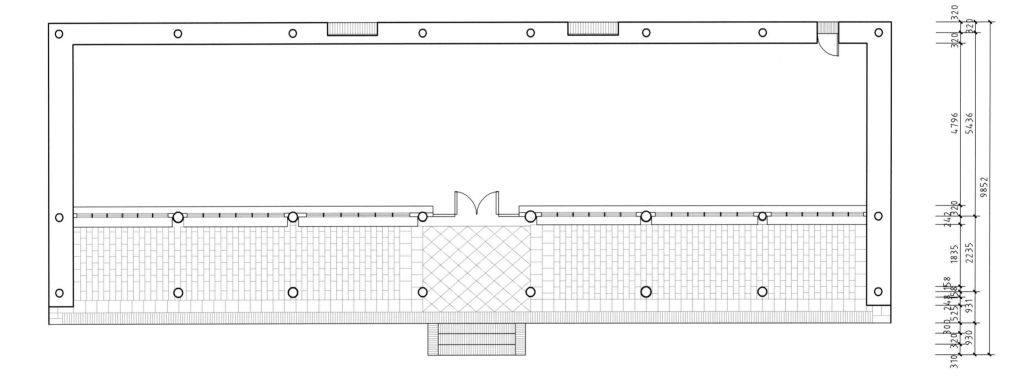

广胜上寺大雄宝殿东配殿平面图

Floor Plan of the Eastern Side Hall of the Treasure Hall of the Great Hero, Upper Guangsheng Monastery

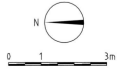

0   1   3m

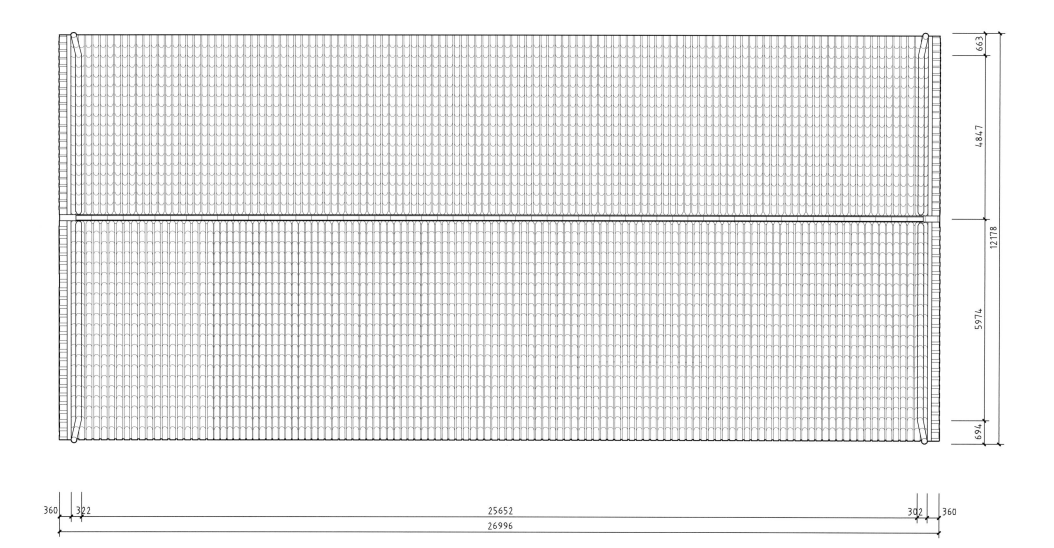

广胜上寺大雄宝殿东配殿屋顶平面图
Top View of the Roof of the Eastern Side Hall of the Treasure Hall of the Great Hero, Upper Guangsheng Monastery

0　1　　3m

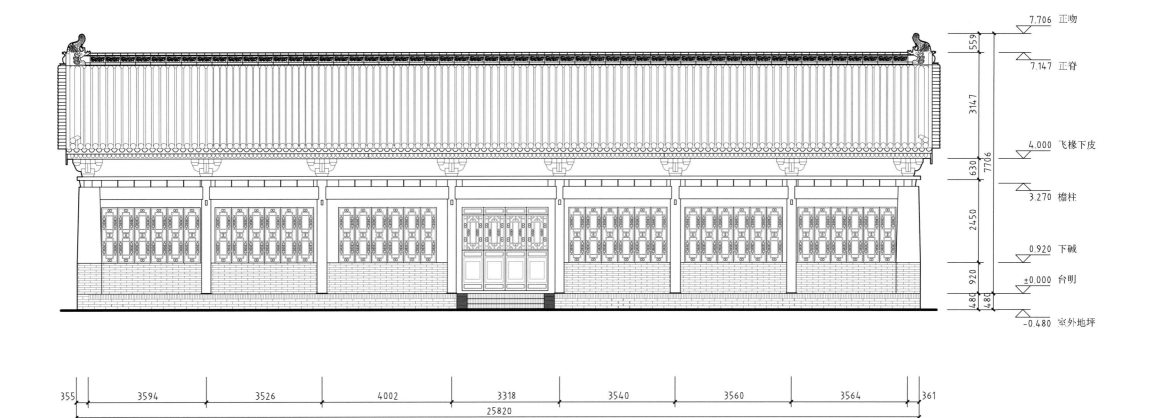

7.706 正吻
7.147 正脊
4.000 飞椽下皮
3.270 檐柱
0.920 下碱
±0.000 台明
-0.480 室外地坪

559
3147
630
2450
920
480
7706
480

355　3594　3526　4002　3318　3540　3560　3564　361
25820

广胜上寺大雄宝殿东配殿正立面图
Front Elevation of the Eastern Side Hall of the Treasure Hall of the Great Hero, Upper Guangsheng Monastery

0　1　3m

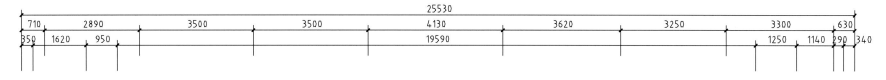

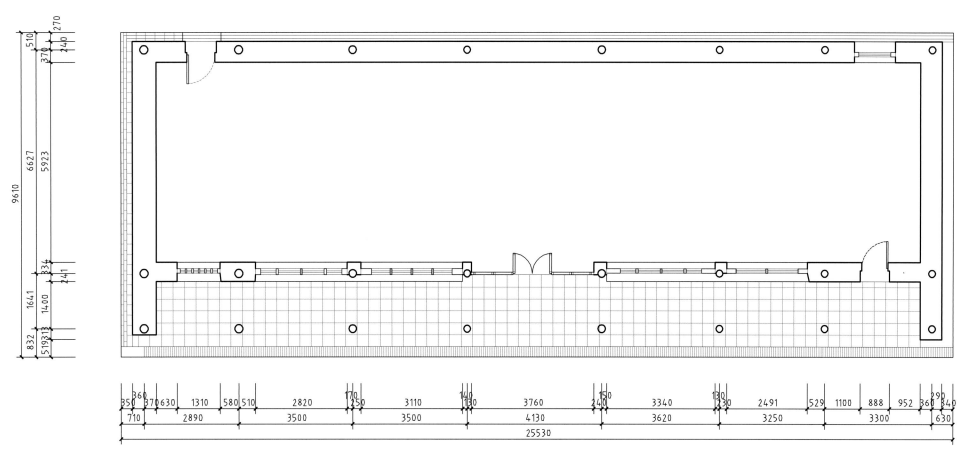

广胜上寺大雄宝殿西配殿平面图

Floor Plan of the Western Side Hall of the Treasure Hall of the Great Hero, Upper Guangsheng Monastery

0   1   3m

N

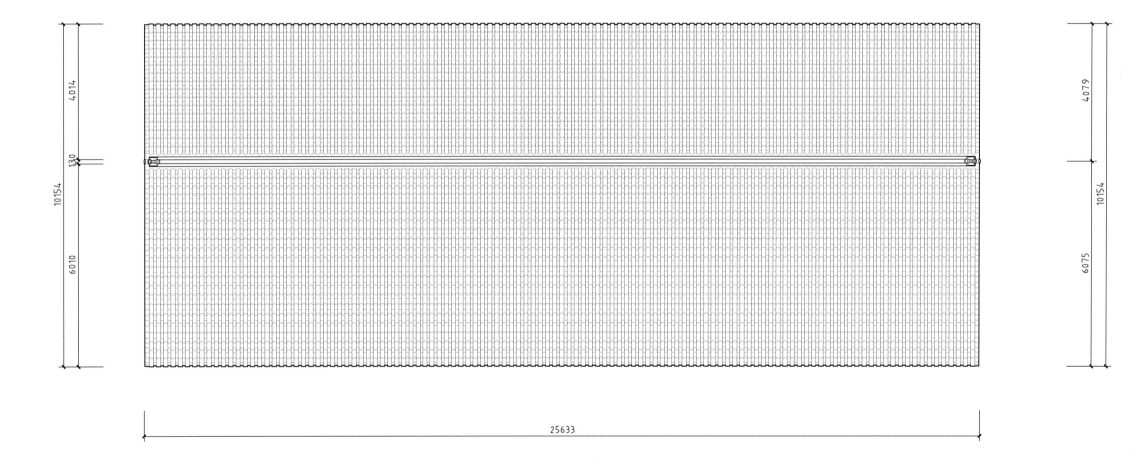

广胜上寺大雄宝殿西配殿屋顶平面图
Top View of the Roof of the Western Side Hall of the Treasure Hall of the Great Hero, Upper Guangsheng Monastery

0　1　　　3m

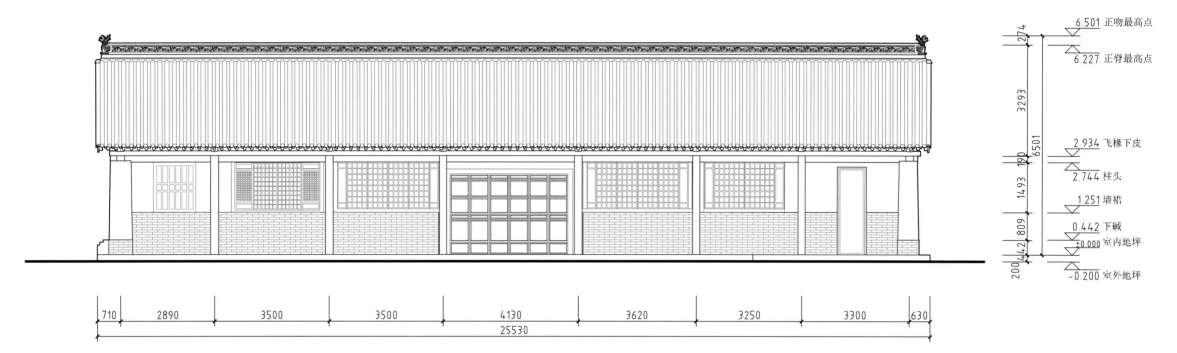

广胜上寺大雄宝殿西配殿正立面图

Front Elevation of the Western Side Hall of the Treasure Hall of the Great Hero, Upper Guangsheng Monastery

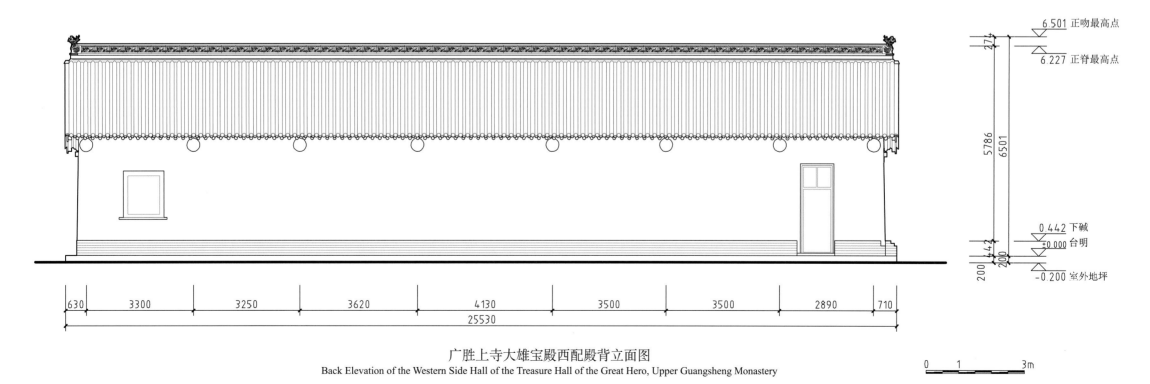

广胜上寺大雄宝殿西配殿背立面图

Back Elevation of the Western Side Hall of the Treasure Hall of the Great Hero, Upper Guangsheng Monastery

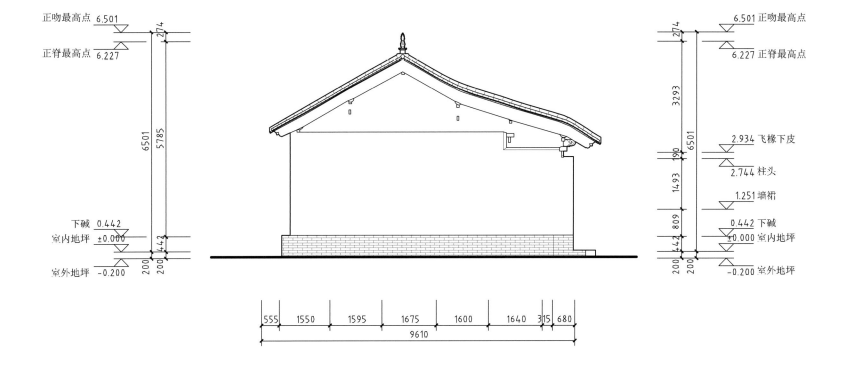

正吻最高点 6.501

正脊最高点 6.227

下碱 0.442
室内地坪 ±0.000
室外地坪 -0.200

2774
6501
5785
442
200 200

6.501 正吻最高点
6.227 正脊最高点

2774
3293
6501

2.934 飞椽下皮
2.744 柱头
1.251 墙裙
0.442 下碱
±0.000 室内地坪
-0.200 室外地坪

190 1493 809 442
200 200

555 1550 1595 1675 1600 1640 315 680
9610

广胜上寺大雄宝殿西配殿侧立面图
Side Elevation of the Western Side Hall of the Treasure Hall of the Great Hero, Upper Guangsheng Monastery

0   1   3m

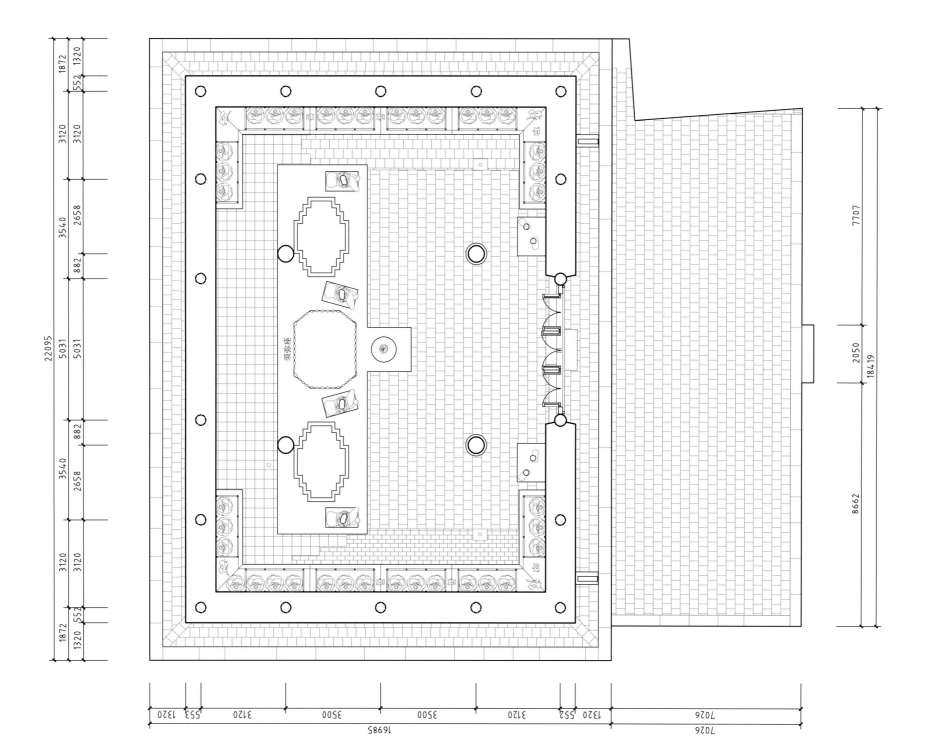

广胜上寺毗卢殿平面图
Floor Plan of Vairocana Hall, Upper Guangsheng Monastery

须弥座

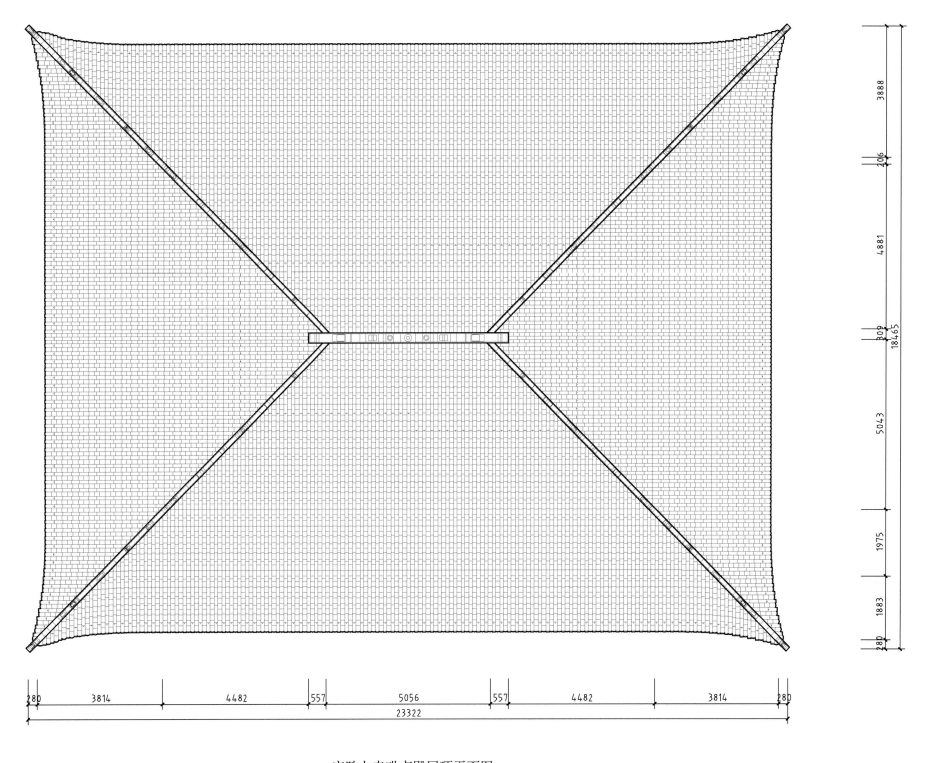

广胜上寺毗卢殿屋顶平面图
Top View of the Roof of Vairocana Hall, Upper Guangsheng Monastery

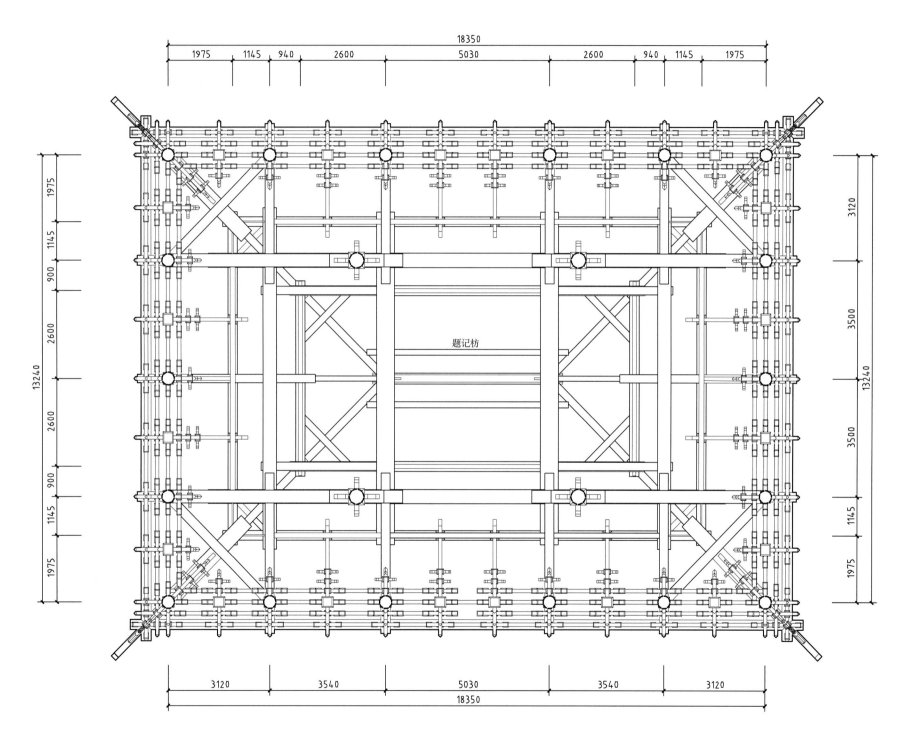

题记枋

广胜上寺毗卢殿梁架仰视平面图
Bottom View of the Truss of Vairocana Hall, Upper Guangsheng Monastery

0　　1　　　　3m

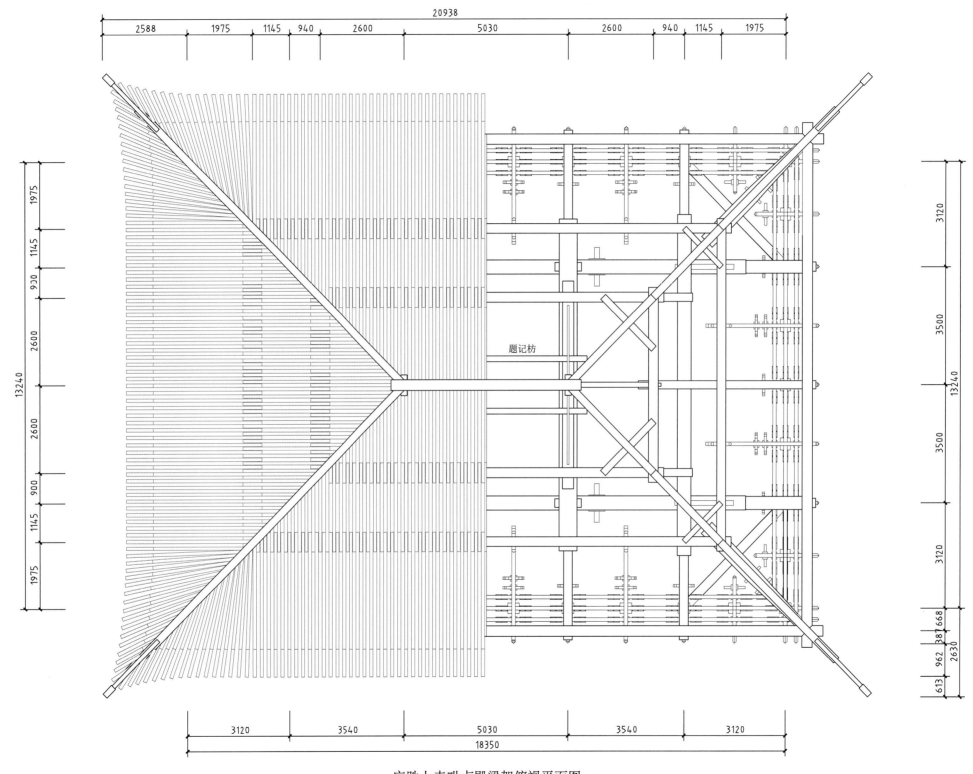

题记枋

广胜上寺毗卢殿梁架俯视平面图
Top View of the Truss of Vairocana Hall, Upper Guangsheng Monastery

0 1 3m

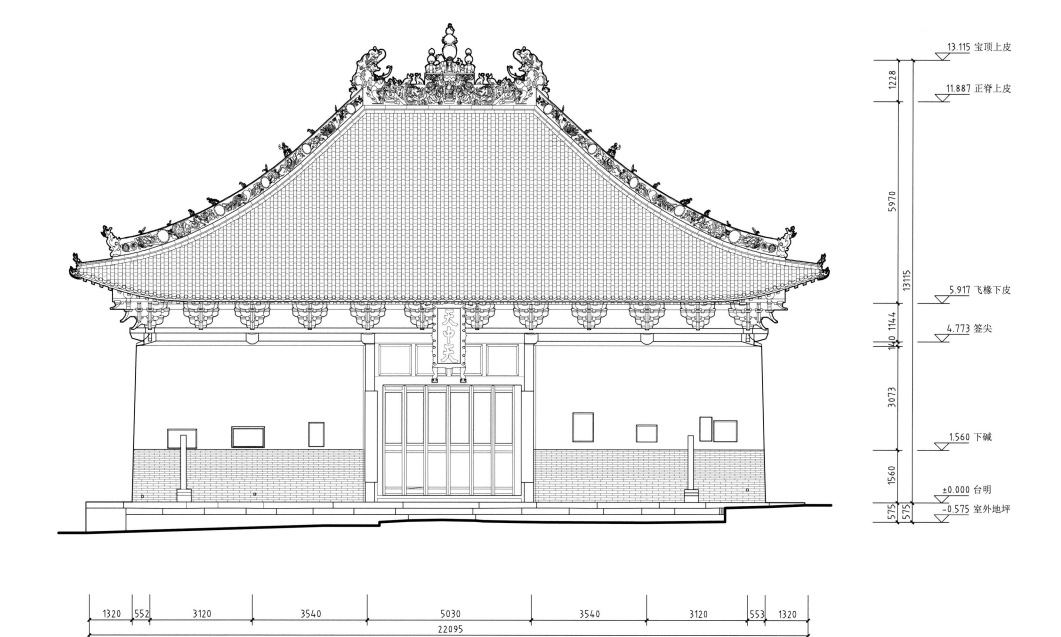

13.115 宝顶上皮

11.887 正脊上皮

5.917 飞椽下皮

4.773 签尖

1.560 下碱

±0.000 台明

-0.575 室外地坪

1228

5970

13115

1144

140

3073

1560

575

575

1320 | 552 | 3120 | 3540 | 5030 | 3540 | 3120 | 553 | 1320

22095

广胜上寺毗卢殿正立面图
Front Elevation of Vairocana Hall, Upper Guangsheng Monastery

0   1   3m

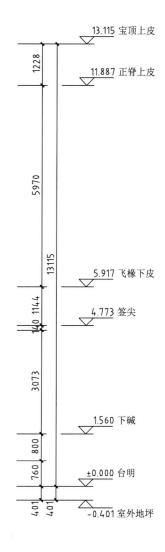

13.115 宝顶上皮

11.887 正脊上皮

5.917 飞椽下皮

4.773 笺尖

1.560 下碱

±0.000 台明

-0.401 室外地坪

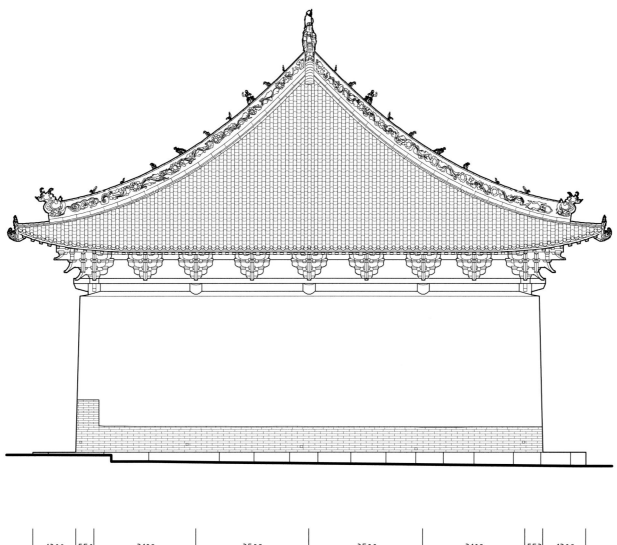

1320 552 3120 3500 3500 3120 553 1320

16985

广胜上寺毗卢殿东立面图
Eastern Elevation of Vairocana Hall, Upper Guangsheng Monastery

0 1 3m

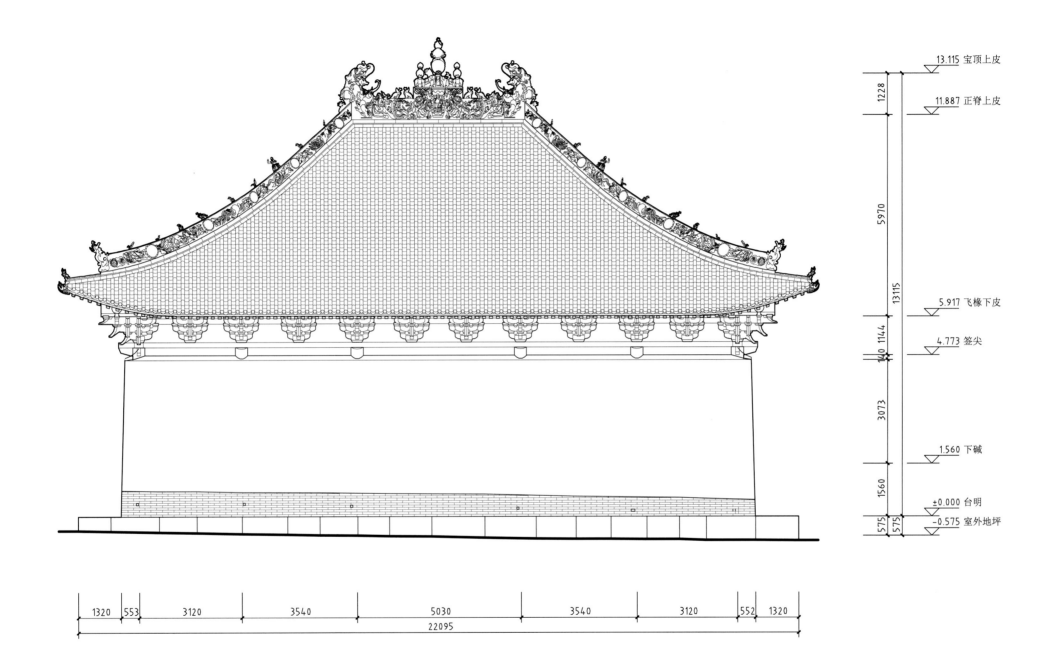

13.115 宝顶上皮

11.887 正脊上皮

5.917 飞椽下皮

4.773 签尖

1.560 下碱

±0.000 台明

-0.575 室外地坪

1228

5970

13115

1144

3073

1560

575

1320 553 3120 3540 5030 3540 3120 552 1320

22095

广胜上寺毗卢殿背立面图

Back Elevation of Vairocana Hall, Upper Guangsheng Monastery

0 1 3m

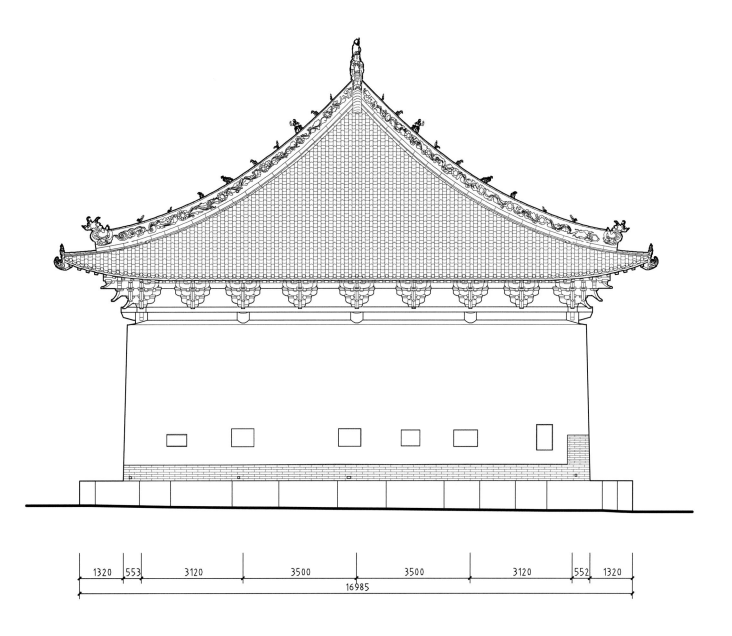

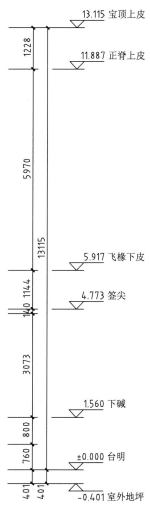

13.115 宝顶上皮

11.887 正脊上皮

5.917 飞椽下皮

4.773 签尖

1.560 下碱

±0.000 台明

-0.401 室外地坪

1228

5970

13115

1144

140

3073

800

760

401

401

1320 | 553 | 3120 | 3500 | 3500 | 3120 | 552 | 1320

16985

广胜上寺毗卢殿西立面图

Western Elevation of Vairocana Hall, Upper Guangsheng Monastery

0　1　3m

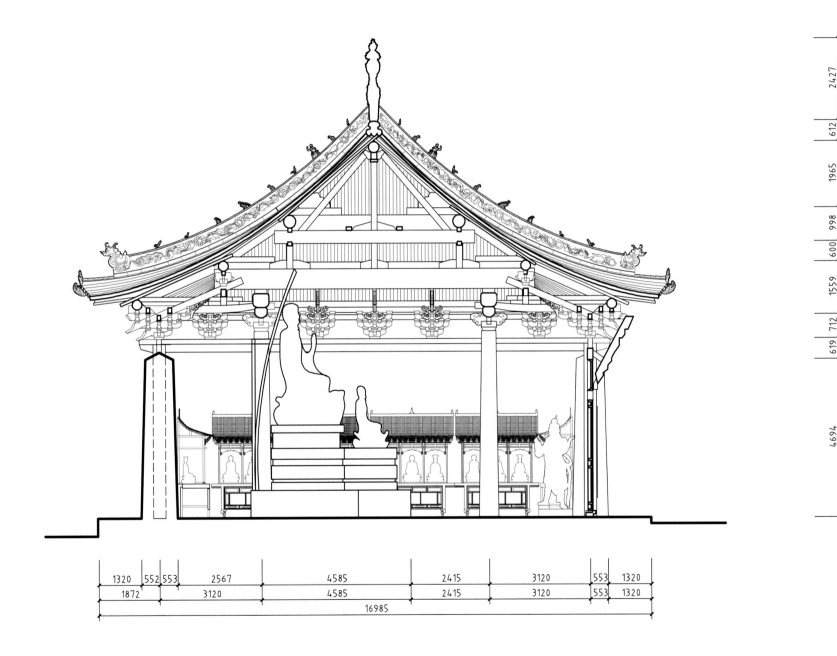

广胜上寺毗卢殿横剖面图
Cross Section of Vairocana Hall, Upper Guangsheng Monastery

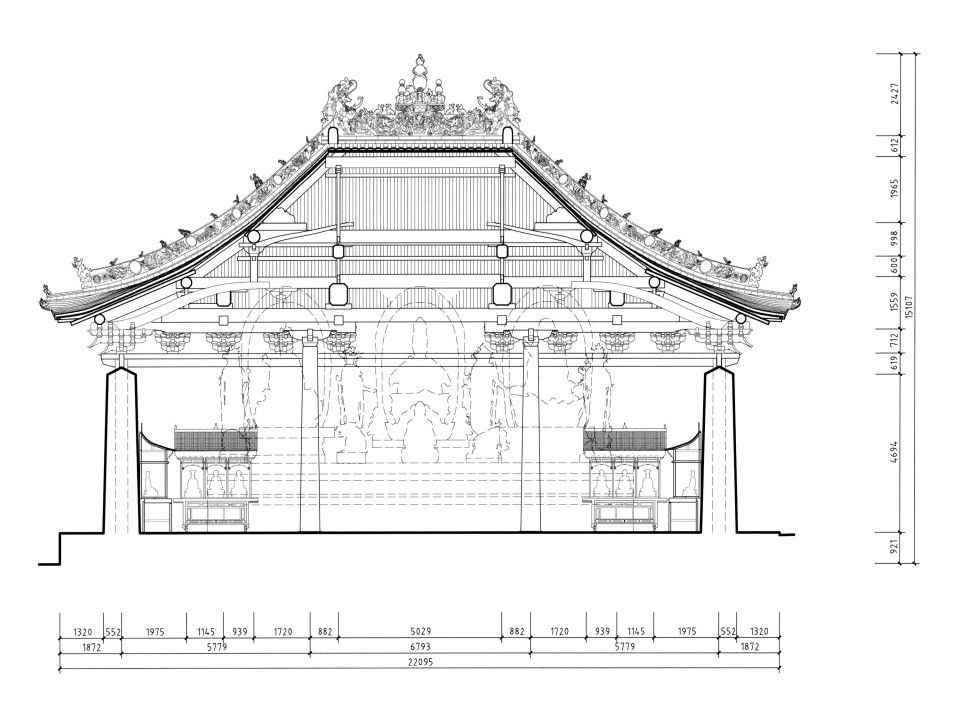

2427

612

1965

998

600

1559

15107

712

619

4694

921

1320 552 1975 1145 939 1720 882 5029 882 1720 939 1145 1975 552 1320

1872 5779 6793 5779 1872

22095

广胜上寺毗卢殿纵剖面图
Longitudinal Section of Vairocana Hall, Upper Guangsheng Monastery

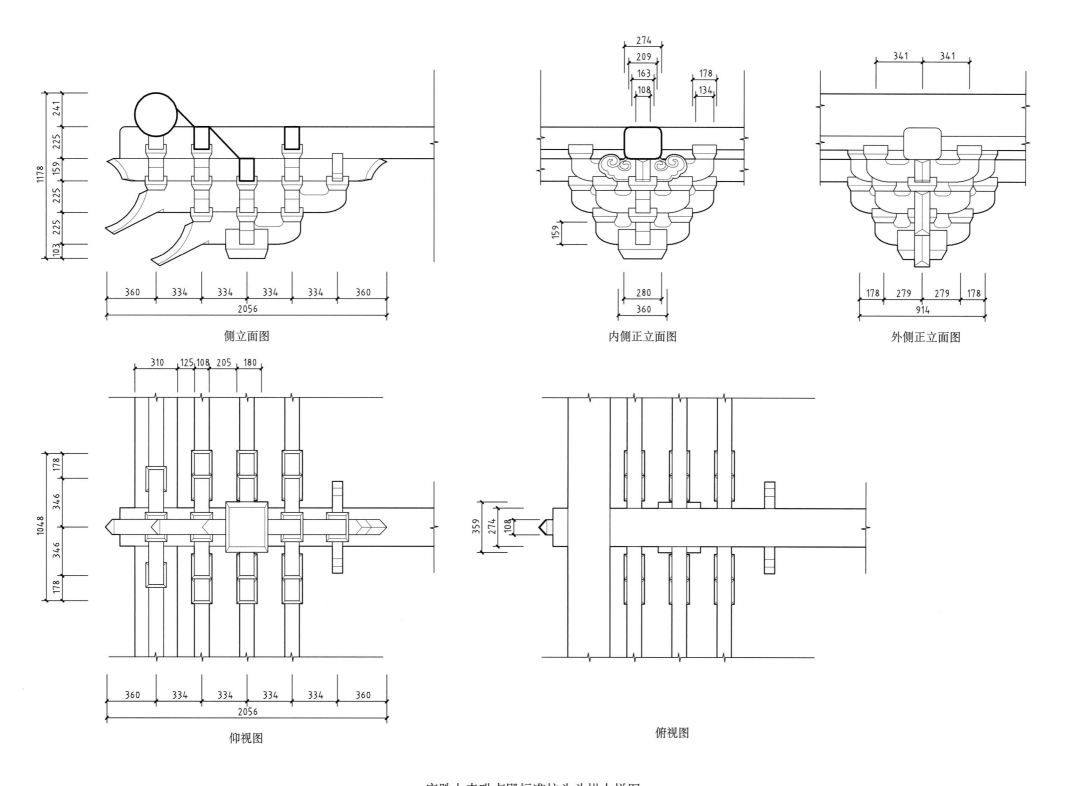

侧立面图

内侧正立面图

外侧正立面图

仰视图

俯视图

广胜上寺毗卢殿标准柱头斗栱大样图

Detailed Drawing of the Standard Bracket Sets on Columns of Vairocana Hall, Upper Guangsheng Monastery

0   0.2   1m

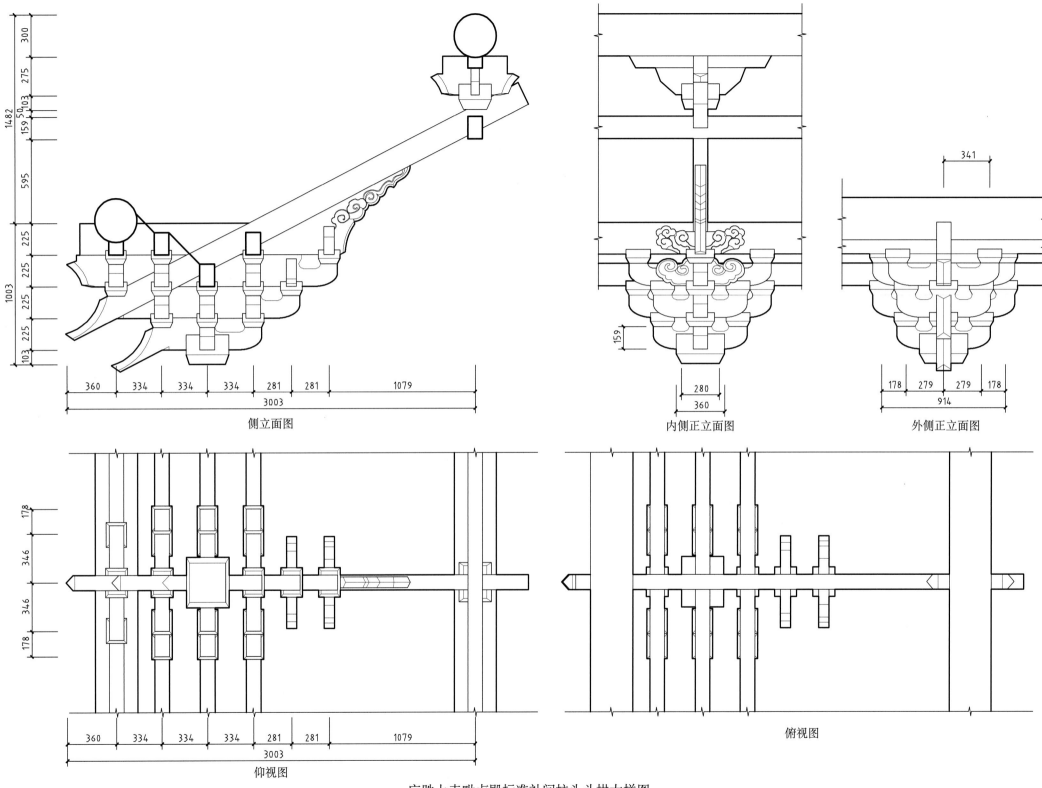

侧立面图

内侧正立面图

外侧正立面图

仰视图

俯视图

广胜上寺毗卢殿标准补间柱头斗栱大样图
Detailed Drawing of the Standard Bracket Sets between Columns of Vairocana Hall, Upper Guangsheng Monastery

0   0.2   1m

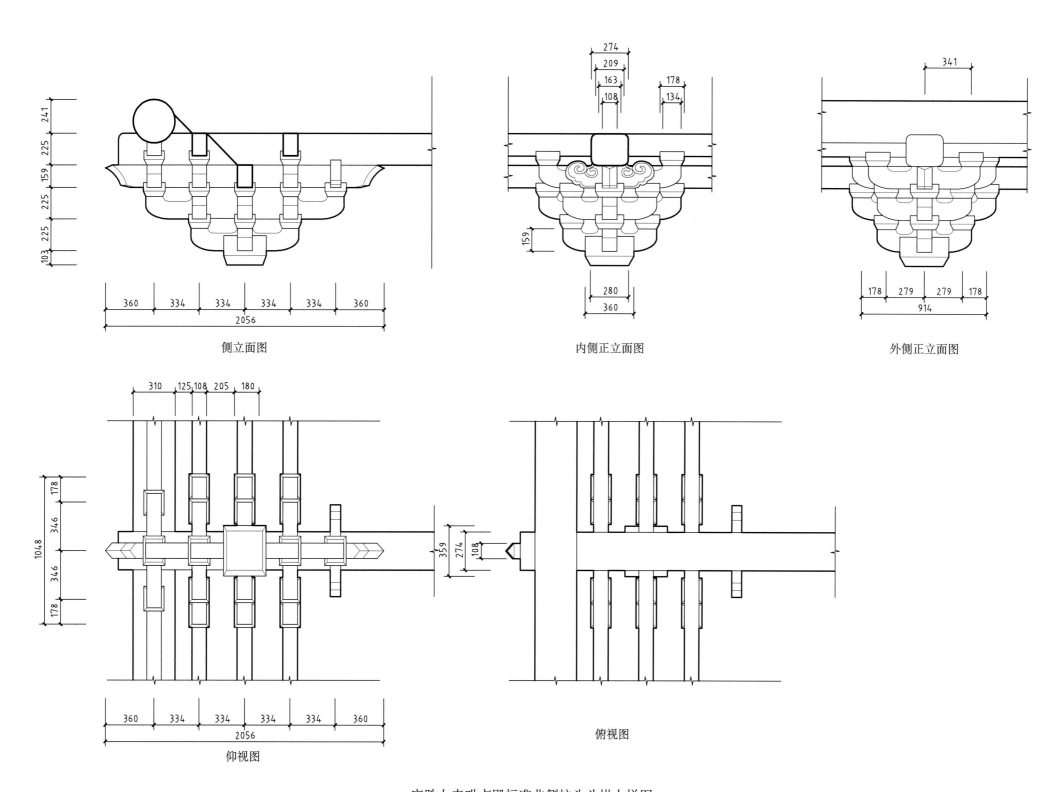

侧立面图   内侧正立面图   外侧正立面图

仰视图   俯视图

广胜上寺毗卢殿标准北侧柱头斗栱大样图

Detailed Drawing of the Standard Bracket Sets on Northern Columns of Vairocana Hall, Upper Guangsheng Monastery

0   0.2   1m

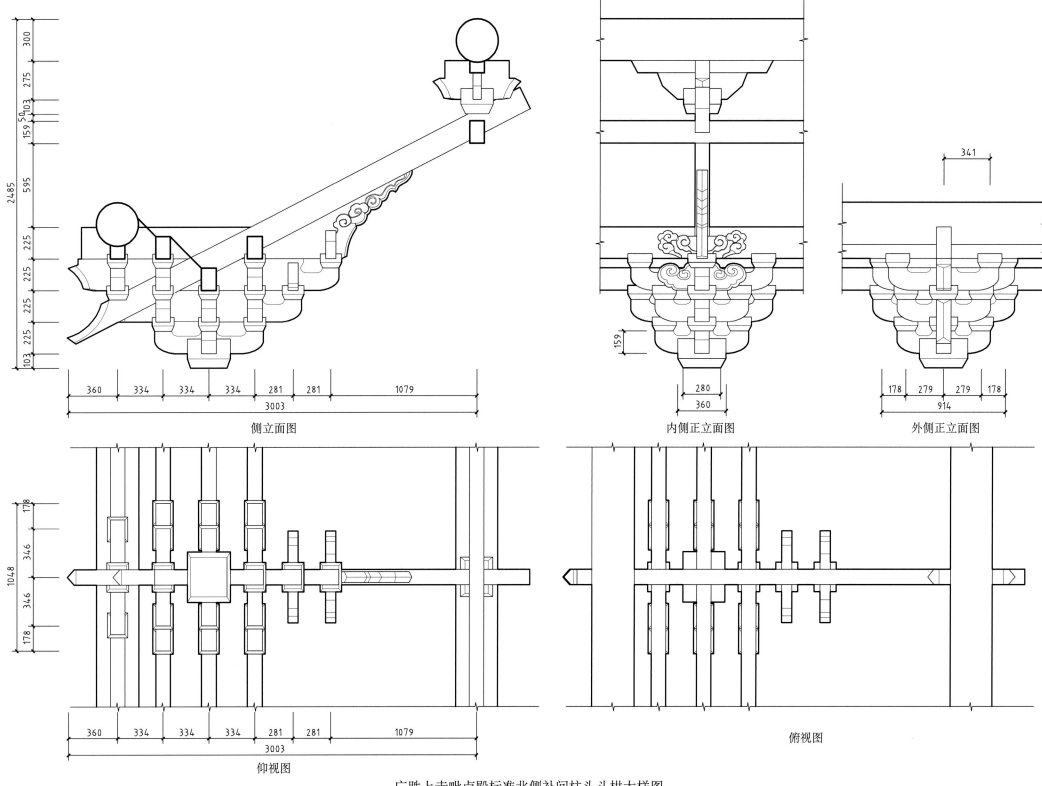

侧立面图

内侧正立面图

外侧正立面图

仰视图

俯视图

广胜上寺毗卢殿标准北侧补间柱头斗栱大样图
Detailed Drawing of the Standard Bracket Sets between Northern Columns of Vairocana Hall, Upper Guangsheng Monastery

0   0.2   1m

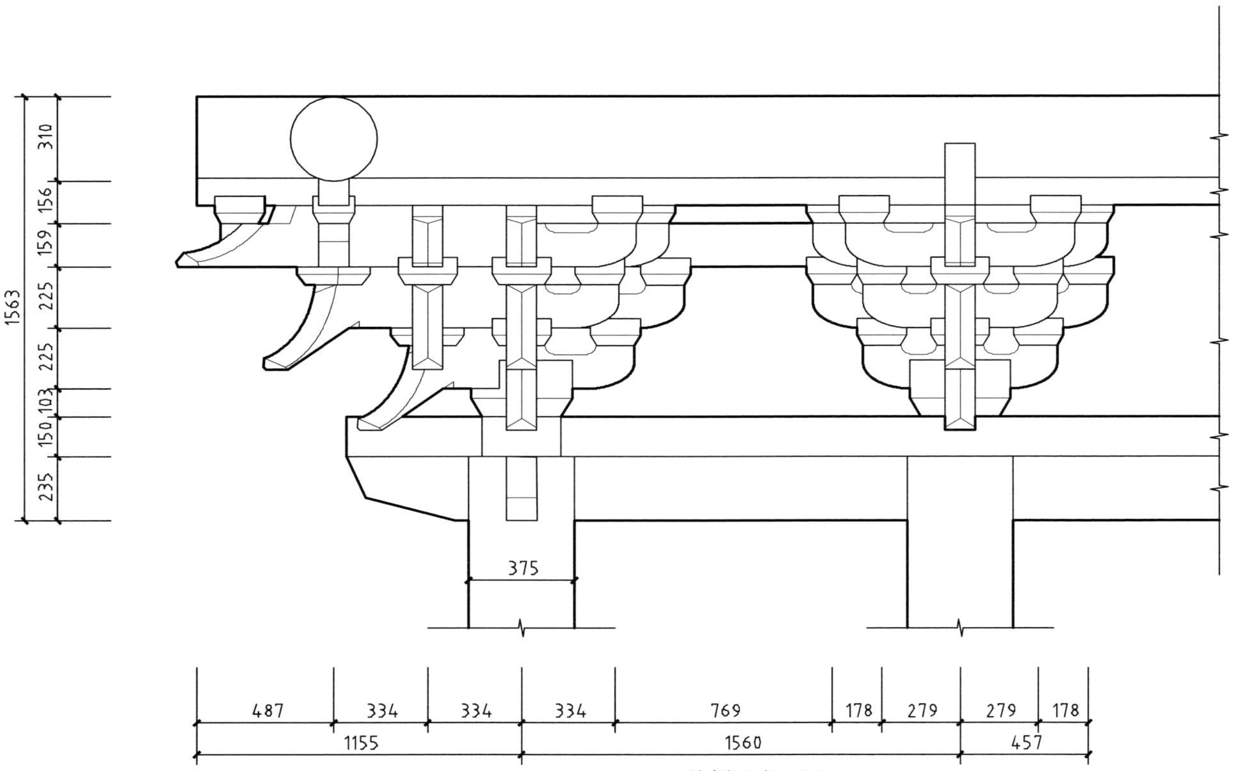

外侧正面图

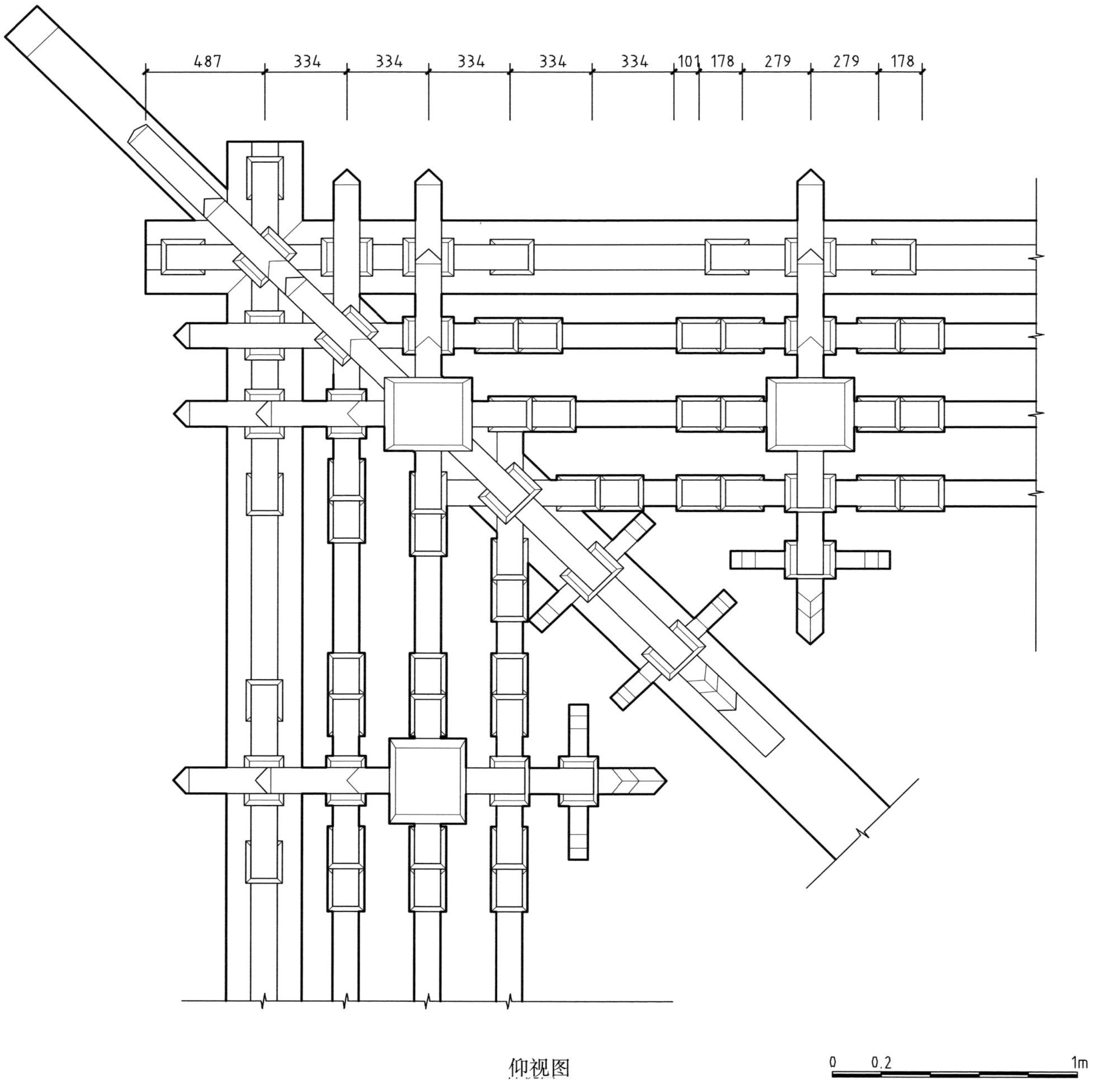

仰视图

广胜上寺毗卢殿后殿转角斗栱大样图（一）

Detailed Drawing of the Bracket Sets on Corners of Vairocana Hall, Upper Guangsheng Monastery (1)

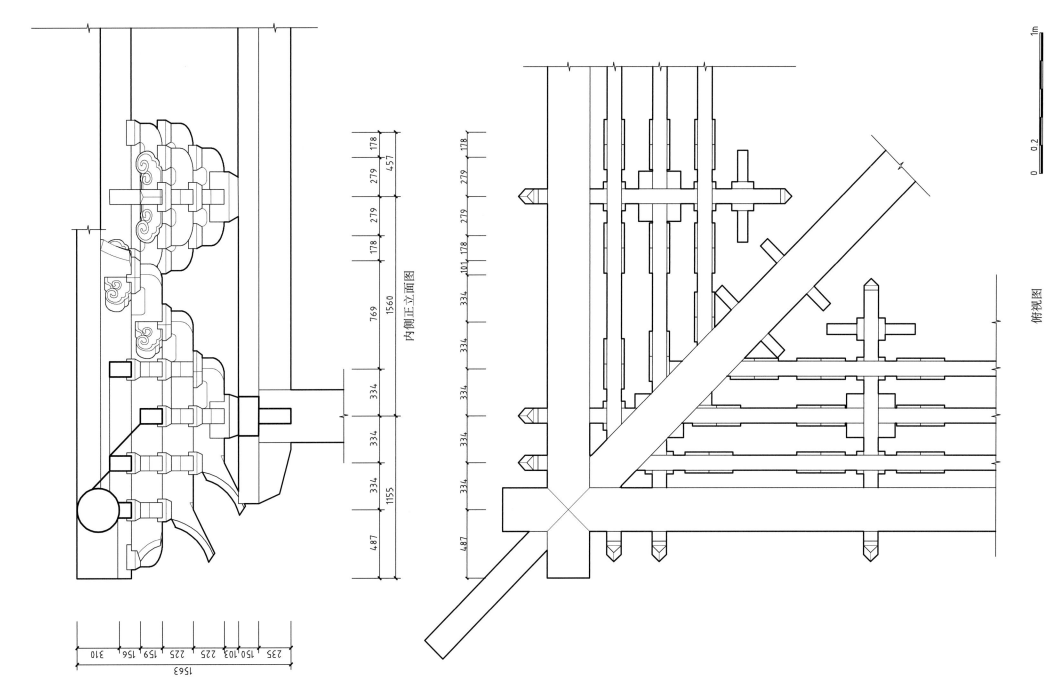

内侧正立面图

俯视图

广胜上寺毗卢遮殿后殿转角斗栱大样图（二）

Detailed Drawing of the Bracket Sets on Corners of Vairocana Hall, Upper Guangsheng Monastery (2)

0   0.2   1m

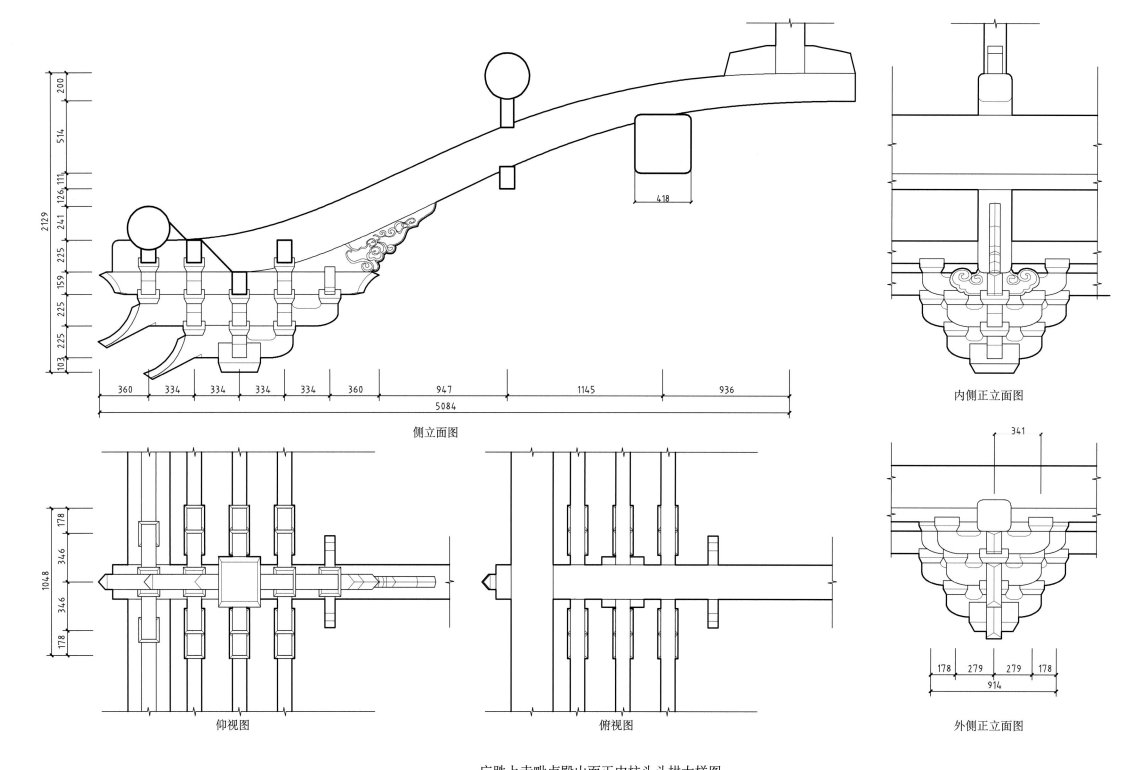

200
514
126,111
126
241
225
159
225
225
103
2129

360 334 334 334 334 360 947 1145 936
5084

侧立面图

418

内侧正立面图

178
346
346
178
1048

341

178 279 279 178
914

仰视图

俯视图

外侧正立面图

广胜上寺毗卢殿山面正中柱头斗栱大样图
Detailed Drawing of the Bracket Sets on the Central Bays of Gable Walls of Vairocana Hall, Upper Guangsheng Monastery

0 0.2 1m

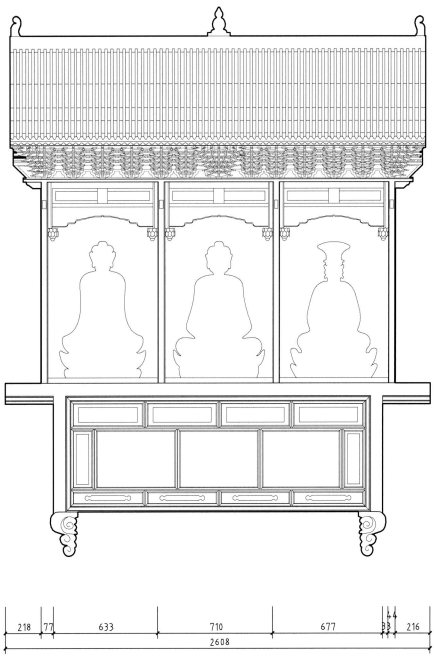

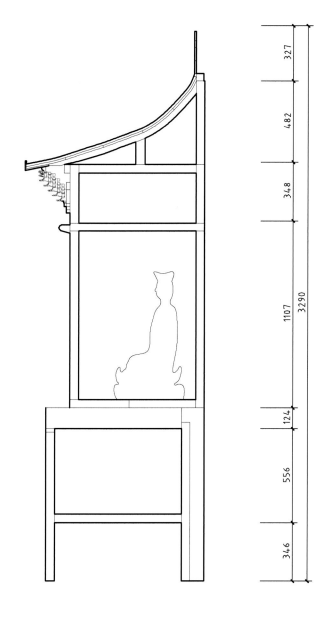

广胜上寺毗卢殿小木作立面图
Elevation of Woodworks in Vairocana Hall, Upper Guangsheng Monastery

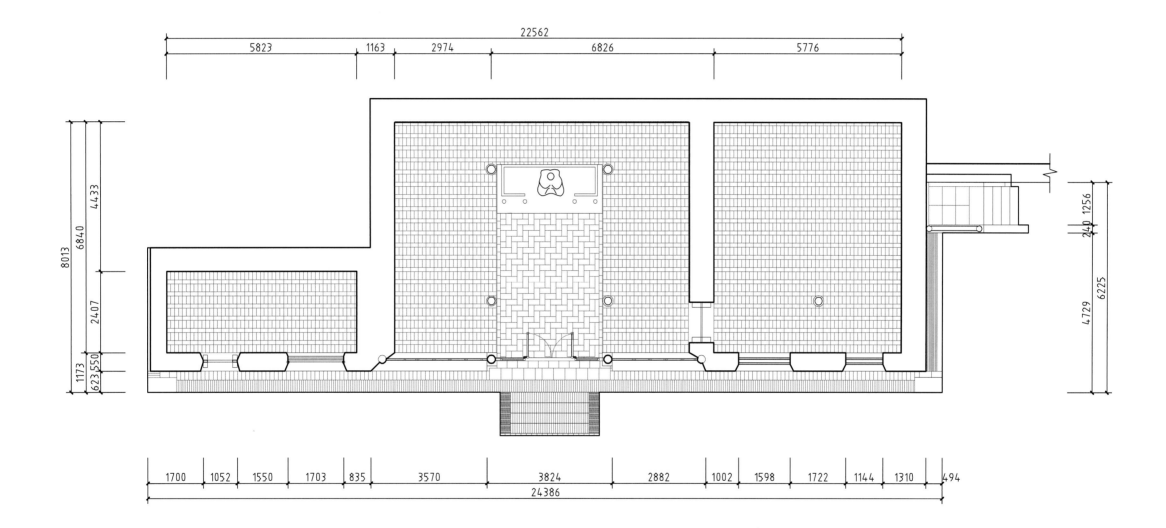

广胜上寺观音殿平面图
Floor Plan of Avalokitesvara Hall, Upper Guangsheng Monastery

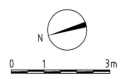

0 1 3m

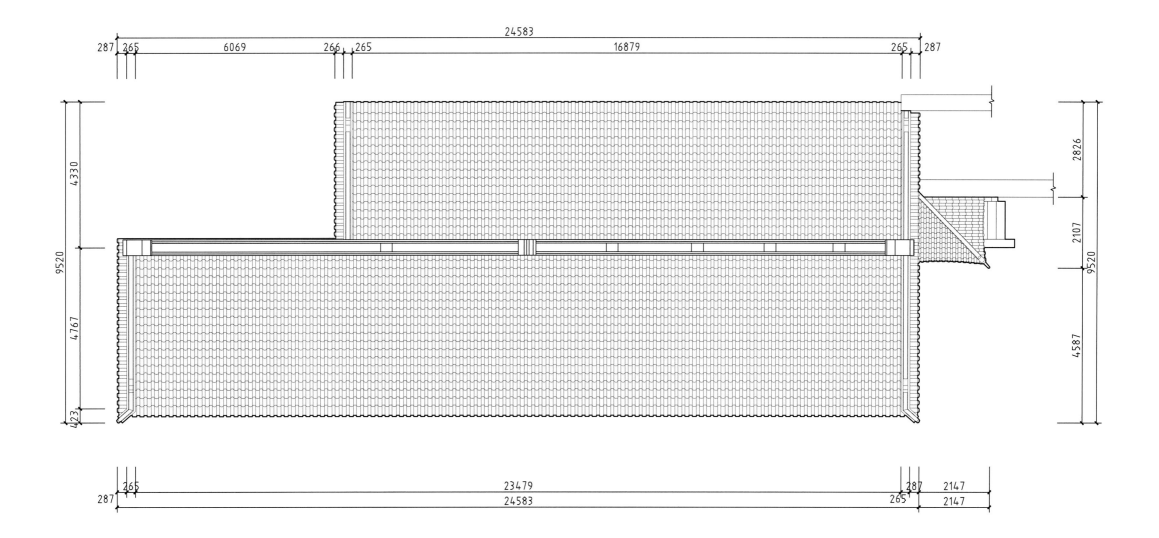

广胜上寺观音殿屋顶平面图
Top View of the Roof of Avalokitesvara Hall, Upper Guangsheng Monastery

0   1   3m

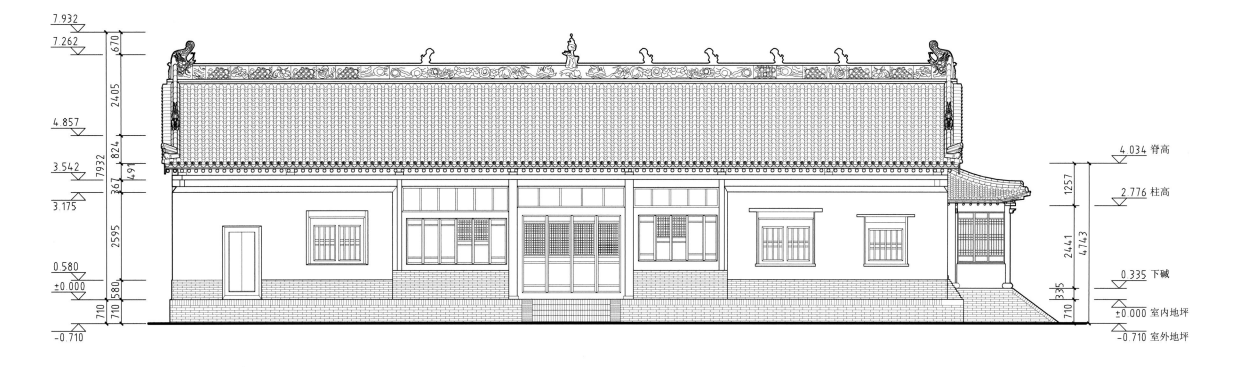

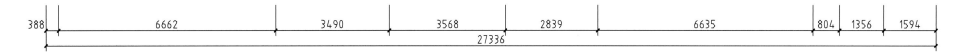

广胜上寺观音殿正立面图
Front Elevation of Avalokitesvara Hall, Upper Guangsheng Monastery

0   1   3m

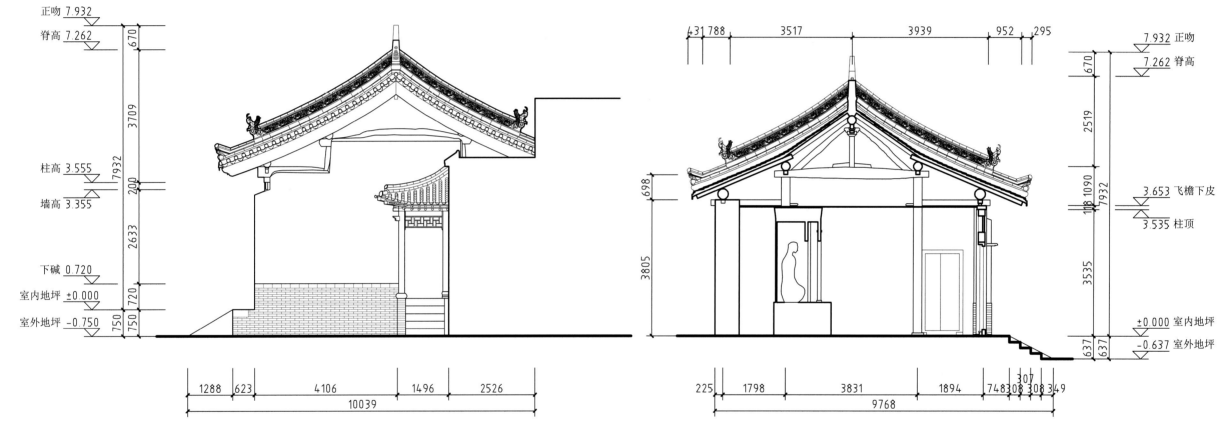

正吻 7.932
脊高 7.262
柱高 3.555
墙高 3.355
下碱 0.720
室内地坪 ±0.000
室外地坪 -0.750

670
3709
7932
200
2633
720
750

1288 | 623 | 4106 | 1496 | 2526
10039

431 788 | 3517 | 3939 | 952 | 295

7.932 正吻
7.262 脊高
3.653 飞檐下皮
3.535 柱顶
±0.000 室内地坪
-0.637 室外地坪

670
2519
1090
118
7932
3535
637

698
3805
637

225 | 1798 | 3831 | 1894 | 748 | 307 308 308 349
9768

广胜上寺观音殿侧立面图
Side Elevation of Avalokitesvara Hall, Upper Guangsheng Monastery

0 1 3m

广胜上寺观音殿横剖面图
Cross Section of Avalokitesvara Hall, Upper Guangsheng Monastery

0 1 3m

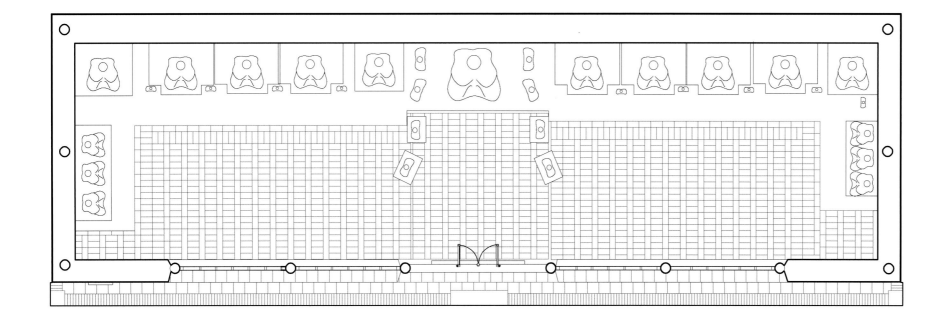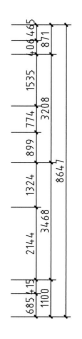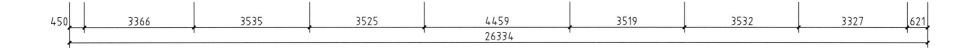

广胜上寺地藏殿平面图
Floor Plan of Ksitigarbha Hall, Upper Guangsheng Monastery

0　1　3m

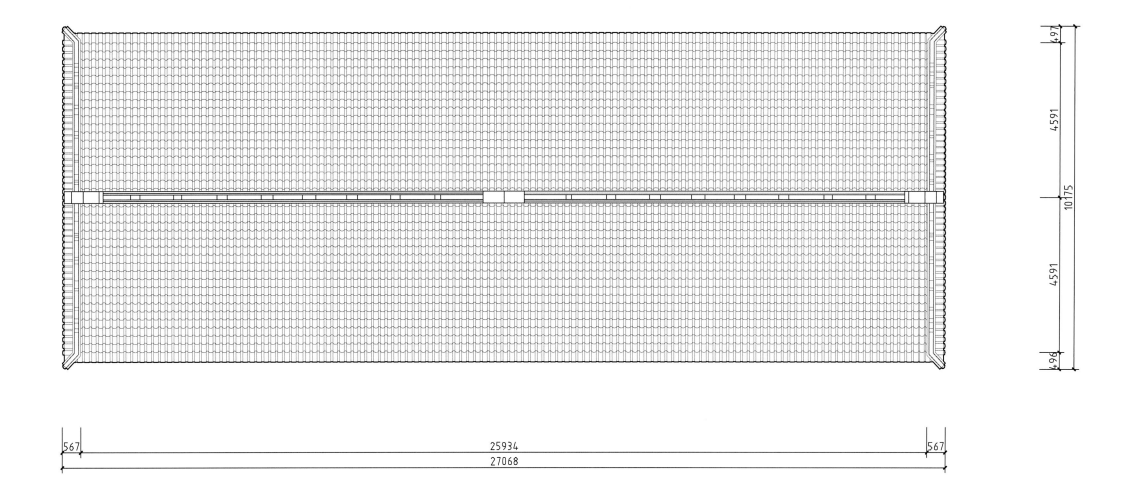

497

4591

10175

4591

496

567   25934   567

27068

广胜上寺地藏殿屋顶平面图
Top View of the Roof of Ksitigarbha Hall, Upper Guangsheng Monastery

0   1   3m

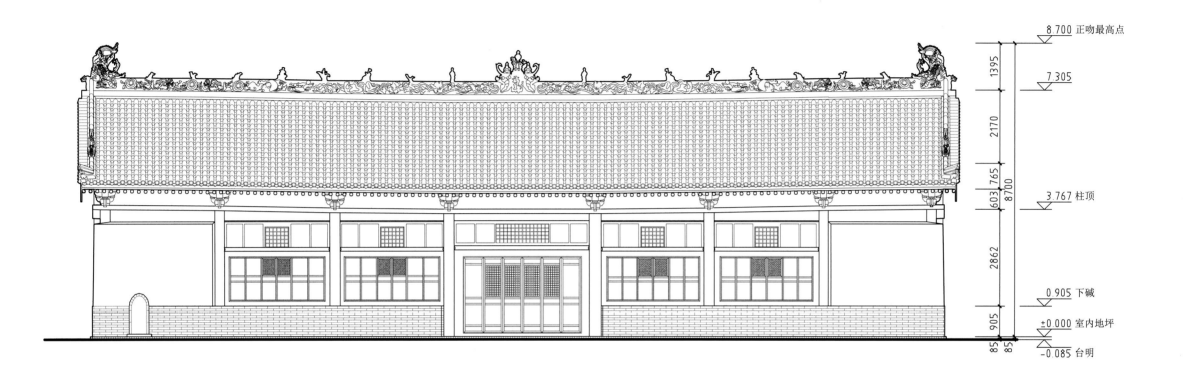

8 700 正吻最高点

7.305

3.767 柱顶

0.905 下碱

±0.000 室内地坪

-0.085 台明

1395

2170

765

603

8700

2862

905

85

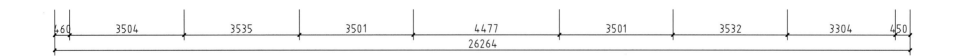

460 3504 3535 3501 4477 3501 3532 3304 450

26264

广胜上寺地藏殿正立面图

Front Elevation of Ksitigarbha Hall, Upper Guangsheng Monastery

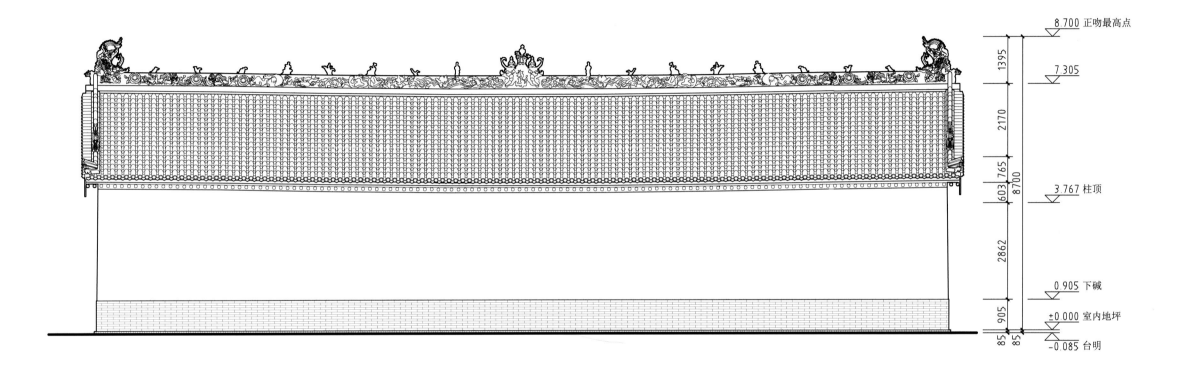

8.700 正吻最高点

7.305

3.767 柱顶

0.905 下碱

±0.000 室内地坪

-0.085 台明

1395

2170

765

603

8700

2862

905

85
85

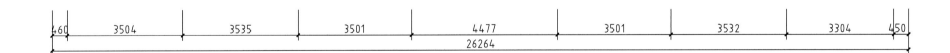

460 3504 3535 3501 4477 3501 3532 3304 450

26264

广胜上寺地藏殿背立面图
Back Elevation of Ksitigarbha Hall, Upper Guangsheng Monastery

0 1 3m

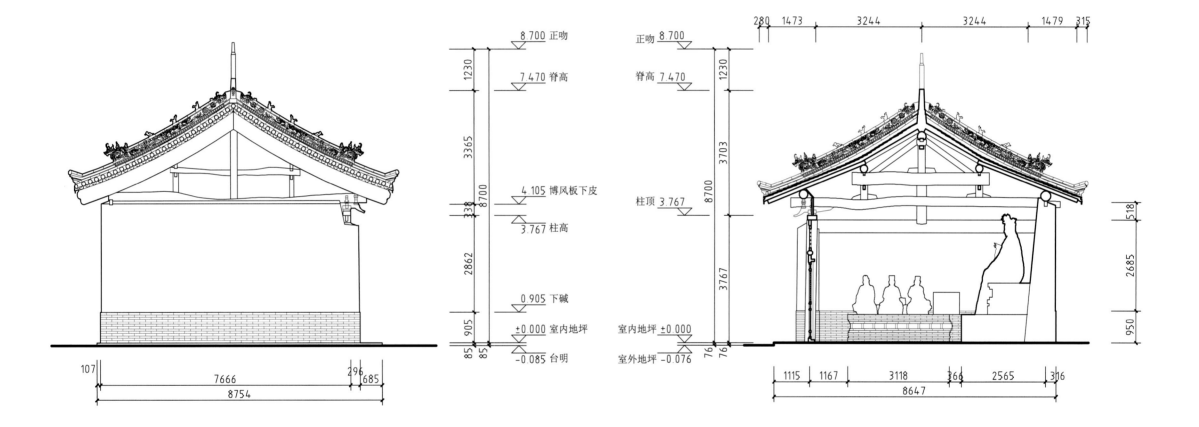

8.700 正吻
1230
7.470 脊高
3365
8700
3365
338
4.105 博风板下皮
3.767 柱高
2862
0.905 下碱
905
±0.000 室内地坪
85
85
-0.085 台明

正吻 8.700
1230
脊高 7.470
3703
8700
柱顶 3.767
3767
室内地坪 ±0.000
76
室外地坪 -0.076
76

280 1473 3244 3244 1479 315

518
2685
950

1115 1167 3118 366 2565 306
8647

107 7666 296 685
8754

广胜上寺地藏殿侧立面图
Side Elevation of Ksitigarbha Hall, Upper Guangsheng Monastery

0 1 3m

广胜上寺地藏殿横剖面图
Cross Section of Ksitigarbha Hall, Upper Guangsheng Monastery

0 1 3m

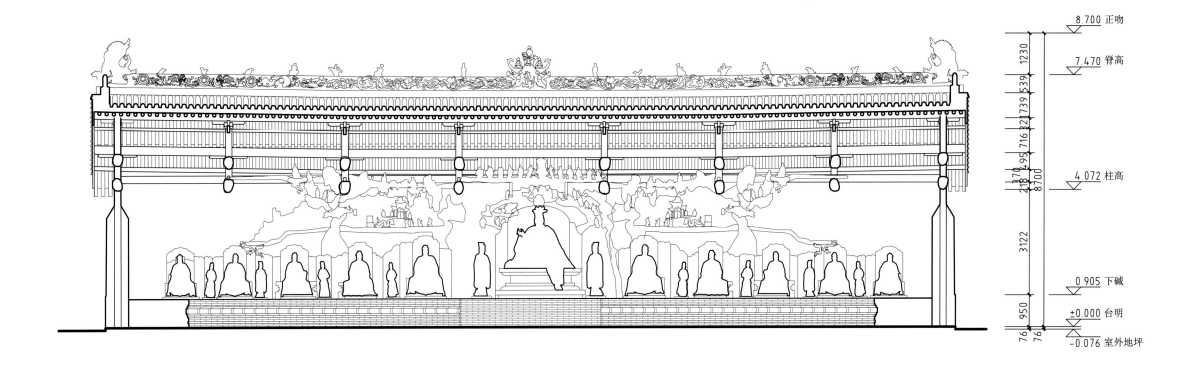

8.700 正吻

7.470 脊高

4.072 柱高

0.905 下碱

±0.000 台明

-0.076 室外地坪

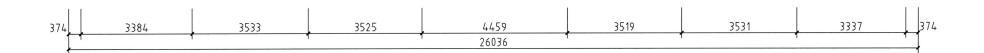

| 374 | 3384 | 3533 | 3525 | 4459 | 3519 | 3531 | 3337 | 374 |

26036

广胜上寺地藏殿纵剖面图
Longitudinal Section of Ksitigarbha Hall, Upper Guangsheng Monastery

0　1　3m

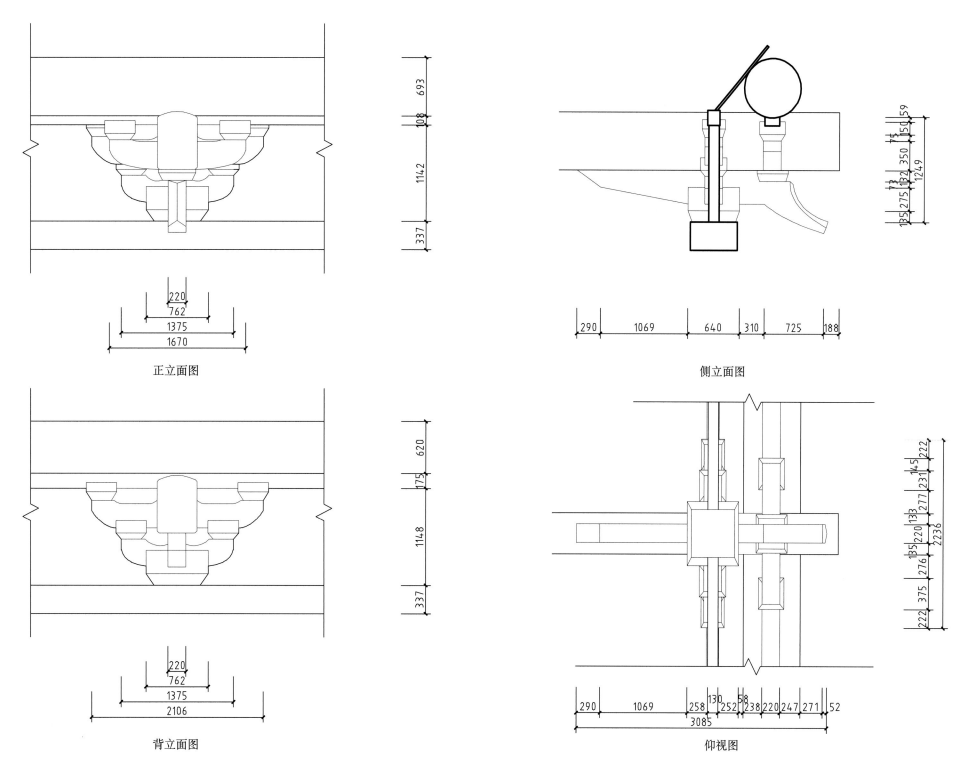

正立面图

侧立面图

背立面图

仰视图

广胜上寺地藏殿斗栱大样图
Detailed Drawing of the Bracket Sets of Ksitigarbha Hall, Upper Guangsheng Monastery

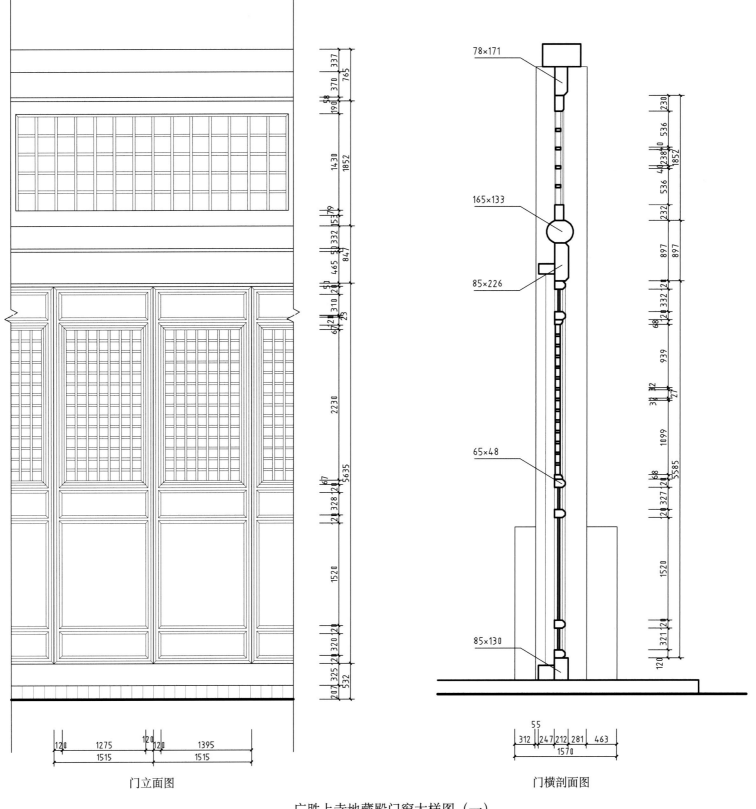

门立面图

门横剖面图

广胜上寺地藏殿门窗大样图（一）
Detailed Drawing of the Doors and Windows of Ksitigarbha Hall, Upper Guangsheng Monastery (1)

0　0.2　　　1m

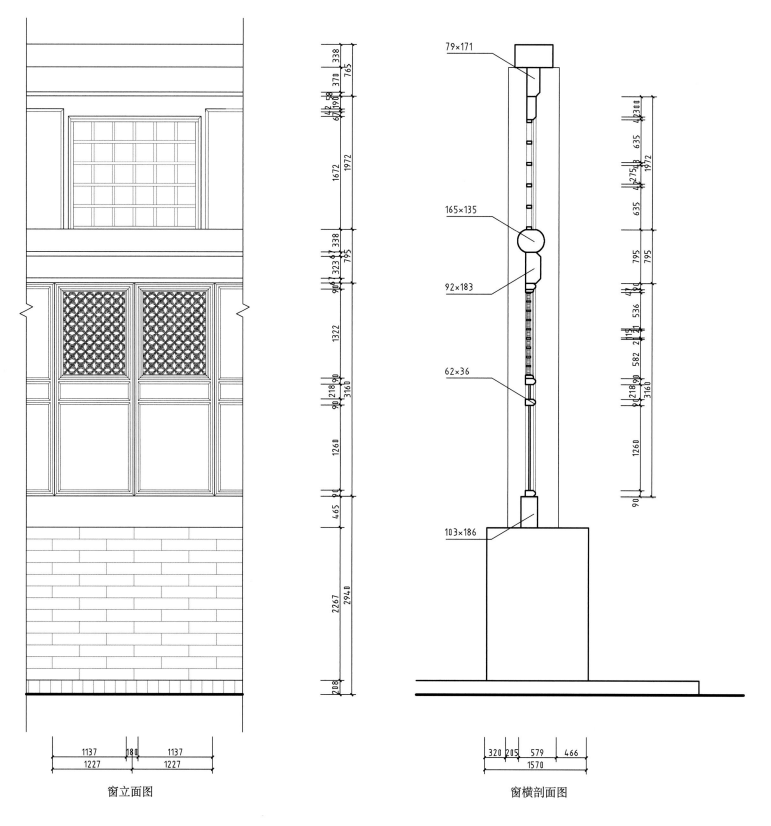

窗立面图

窗横剖面图

广胜上寺地藏殿门窗大样图（二）
Detailed Drawing of the Doors and Windows of Ksitigarbha Hall, Upper Guangsheng Monastery (2)

水神庙

Water God's Temple

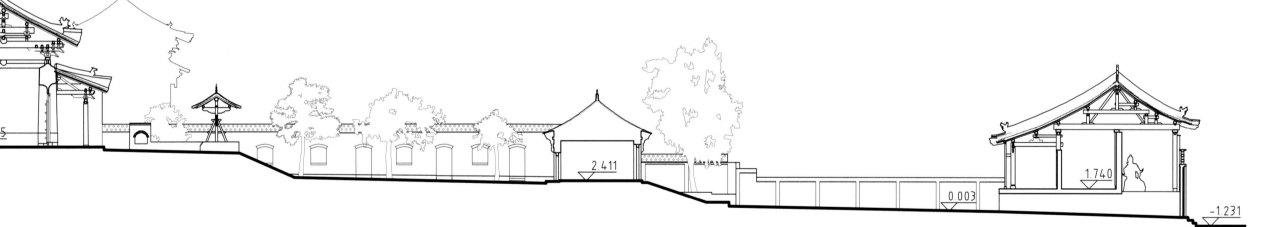

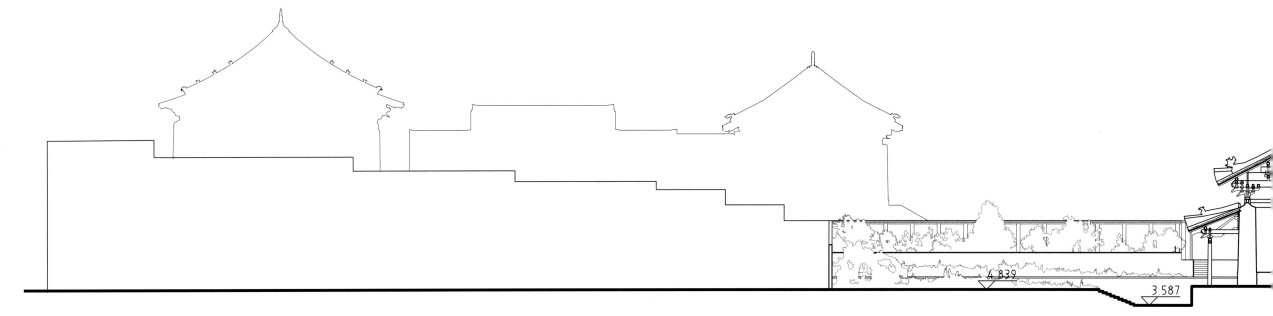

4.839

3.587

水神庙院落总纵剖面图
Overall Longitudinal Section of the Water God's Temple

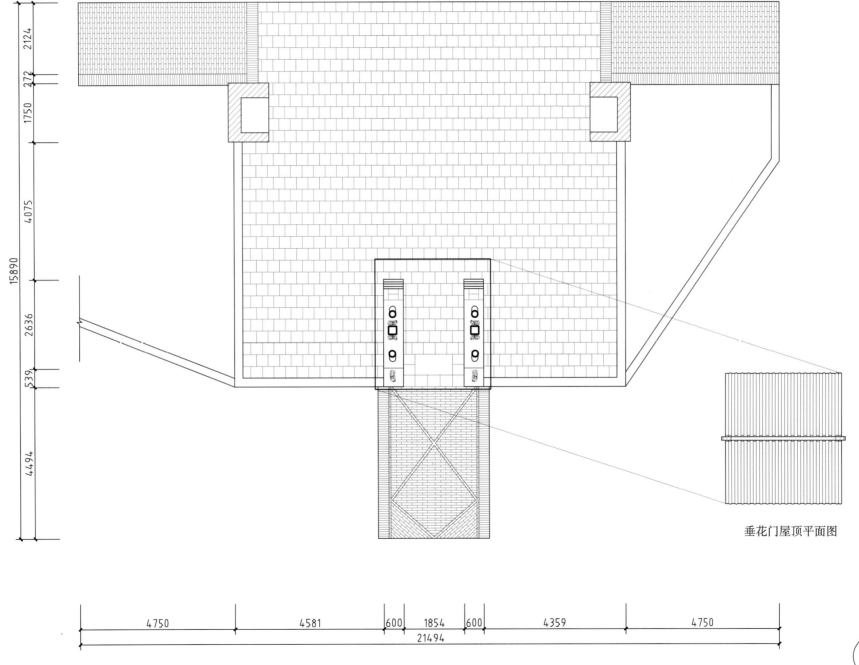

垂花门屋顶平面图

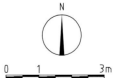

N

水神庙垂花门平面图
Floor Plan of the Floral-pendant Gate, Water God's Temple

0　1　　3m

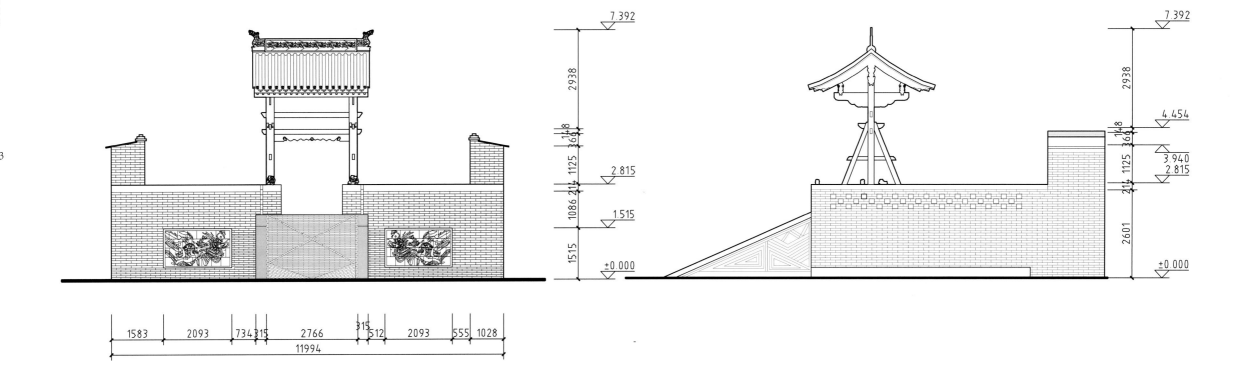

水神庙垂花门正立面图
Front Elevation of the Floral-pendant Gate, Water God's Temple

水神庙垂花门侧立面图
Side Elevation of the Floral-pendant Gate, Water God's Temple

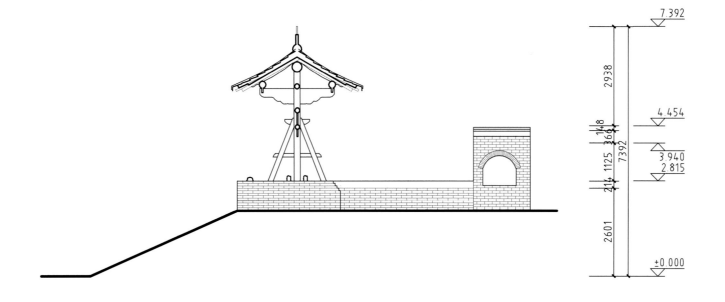

7.392

2938

1148

366

7392

4.454

214 1125

3.940

2.815

214

2601

+0.000

水神庙垂花门横剖面图
Cross Section of the Floral-pendant Gate, Water God's Temple

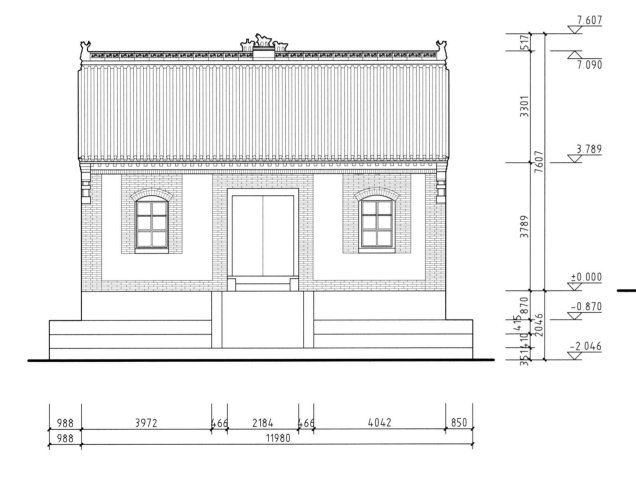

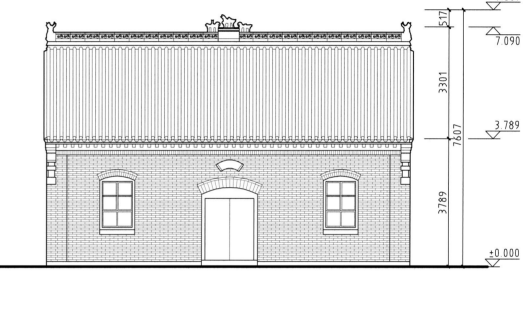

水神庙山门正立面图
Front Elevation of the Monastery Gate, Water God's Temple

水神庙山门背立面图
Back Elevation of the Monastery Gate, Water God's Temple

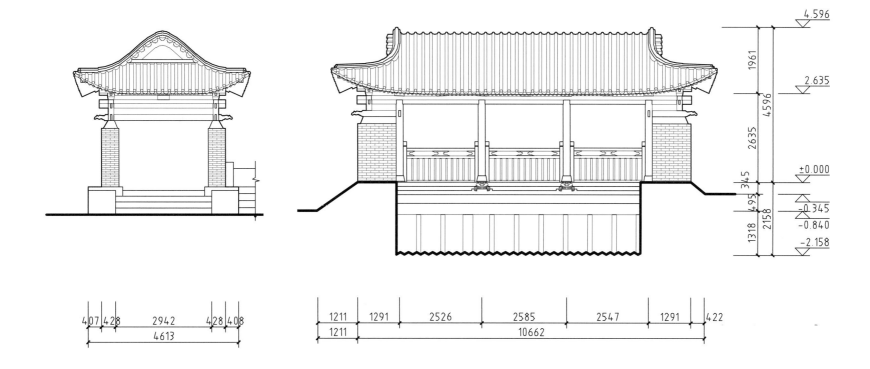

水神庙分水亭侧立面图
Side Elevation of the Pavillion of Water Allocation, Water God's Temple

水神庙分水亭正立面图
Front Elevation of the Pavillion of Water Allocation, Water God's Temple

0   1   3m

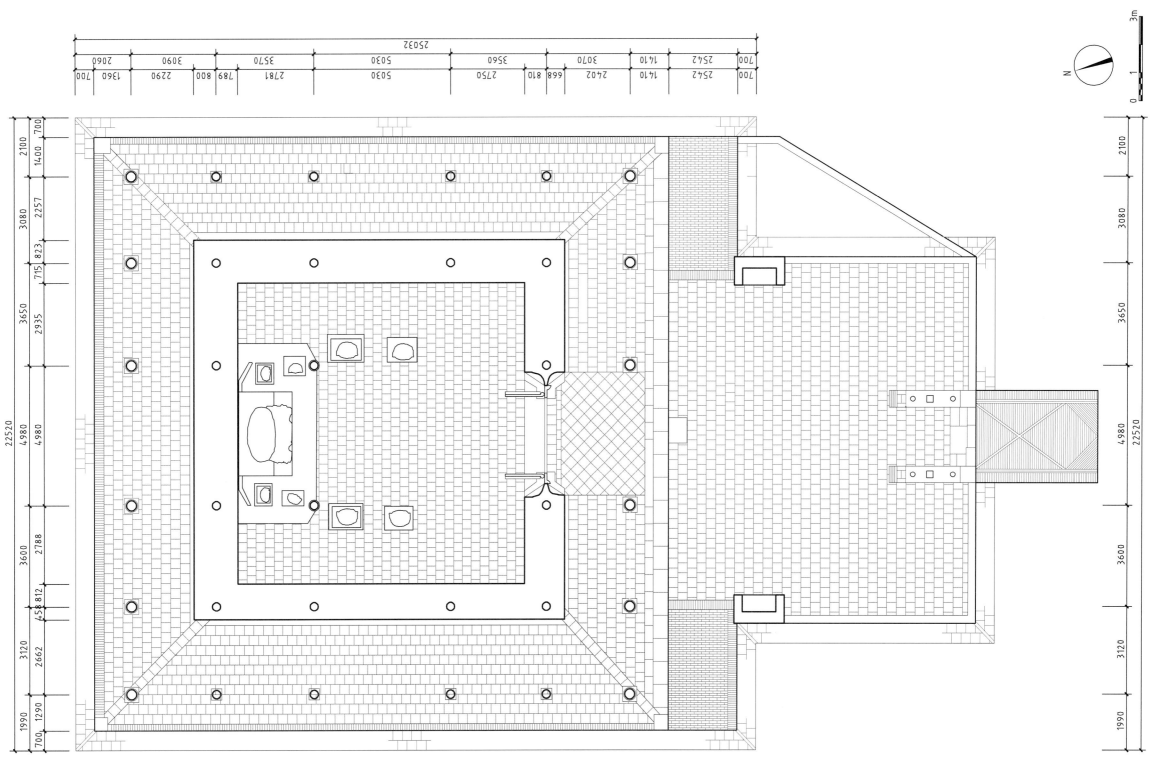

水神庙明应王殿平面图

Floor Plan of the Mingying Wang Hall, Water God's Temple

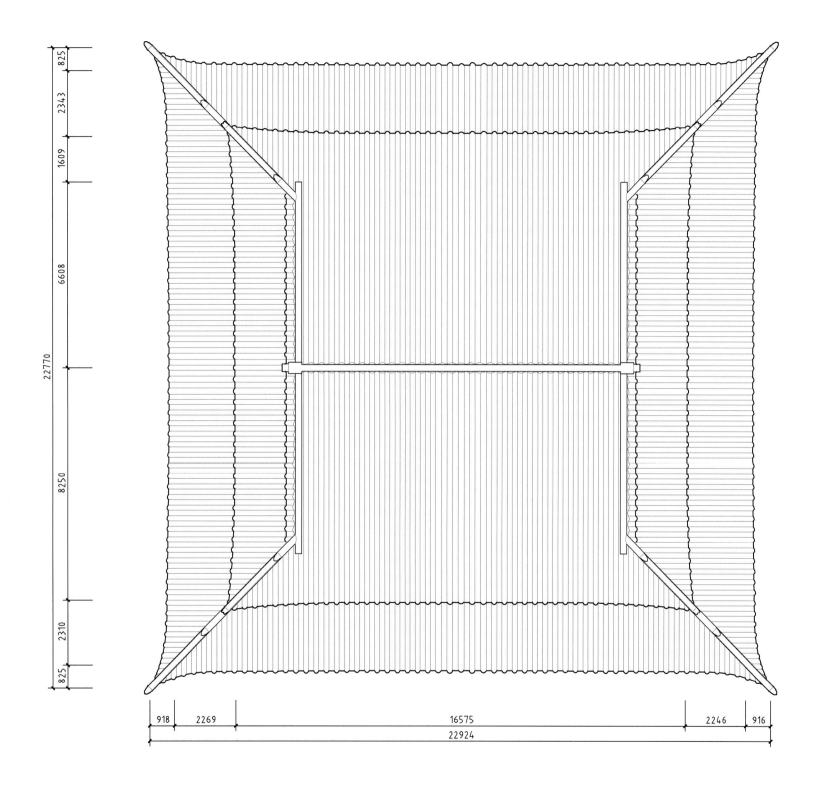

水神庙明应王殿屋顶平面图
Top View of the Roof of the Mingying Wang Hall, Water God's Temple

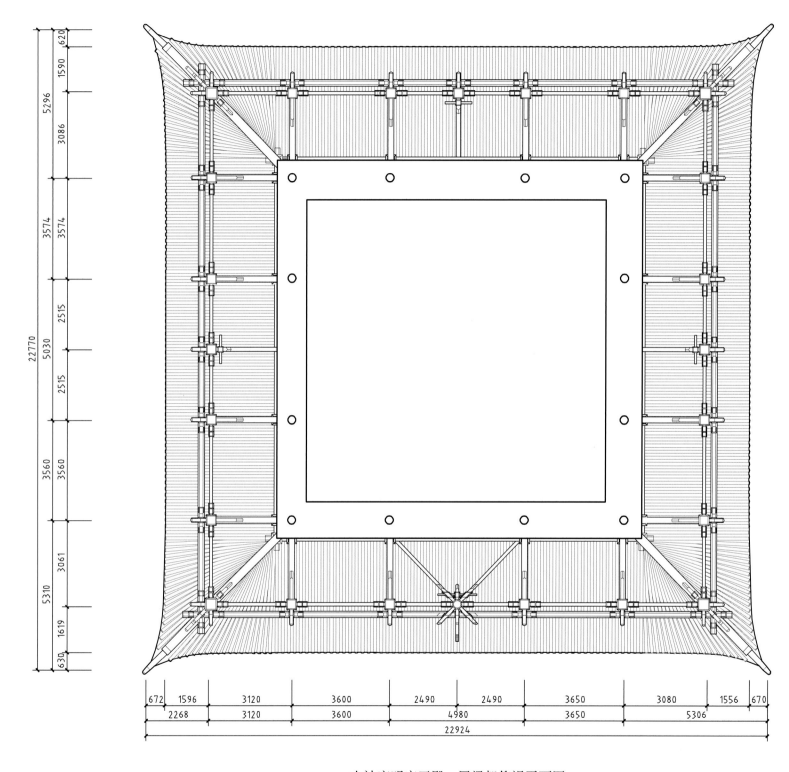

水神庙明应王殿一层梁架仰视平面图
Bottom View of the Truss of the Ground Floor of the Mingying Wang Hall, Water God's Temple

0   1   3m

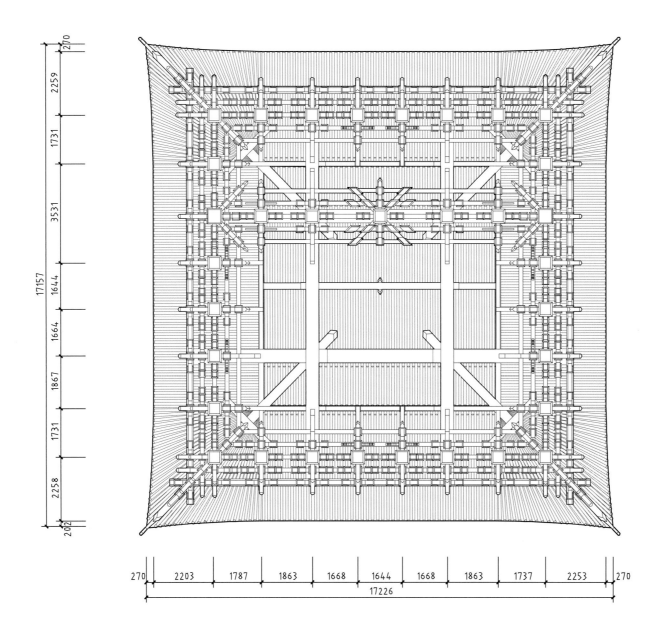

水神庙明应王殿二层梁架仰视平面图
Bottom View of the Truss of the Second Floor of the Mingying Wang Hall, Water God's Temple

0 1 3m

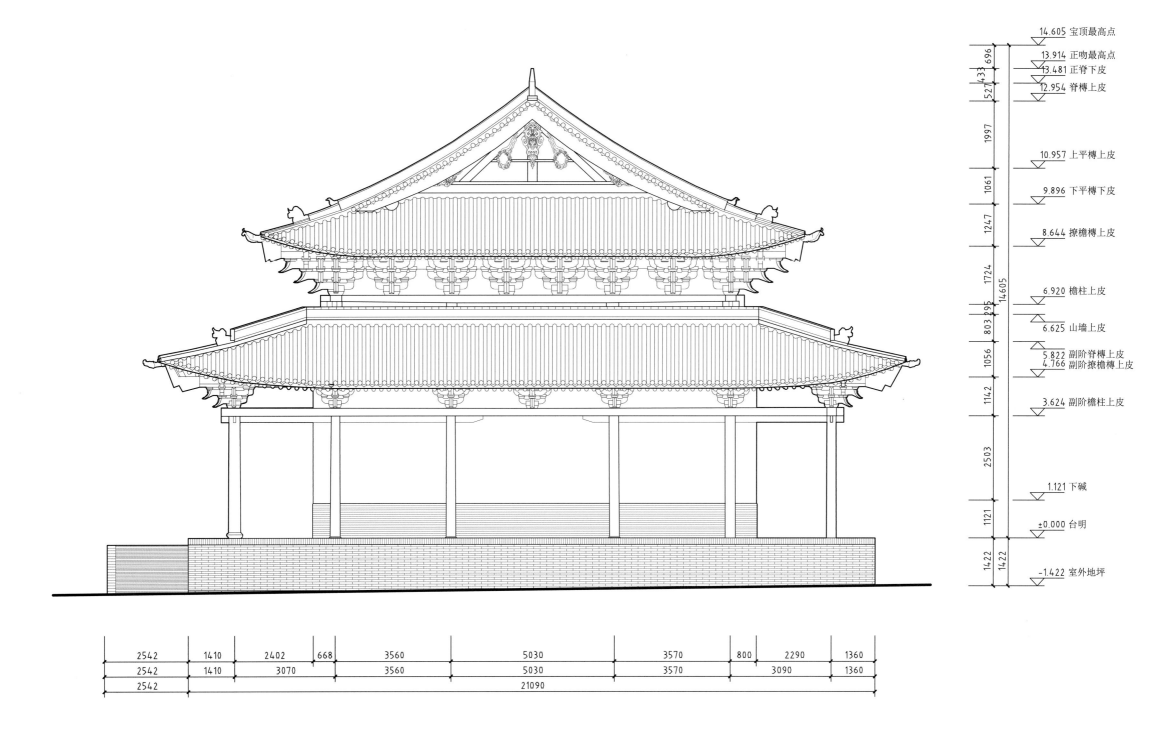

14.605 宝顶最高点
13.914 正吻最高点
13.481 正脊下皮
12.954 脊槫上皮

10.957 上平槫上皮
9.896 下平槫下皮
8.644 撩檐槫上皮

6.920 檐柱上皮
6.625 山墙上皮
5.822 副阶脊槫上皮
4.766 副阶撩檐槫上皮
3.624 副阶檐柱上皮

1.121 下碱
±0.000 台明
-1.422 室外地坪

696
433
527
1997
1061
1247
1724
395
803
1056
1142
2503
1121
14.22

14.605
14.22

2542 1410 2402 668 3560 5030 3570 800 2290 1360
2542 1410 3070 3560 5030 3570 3090 1360
2542 21090

水神庙明应王殿侧立面图
Side Elevation of the Mingying Wang Hall, Water God's Temple

0   1   3m

132

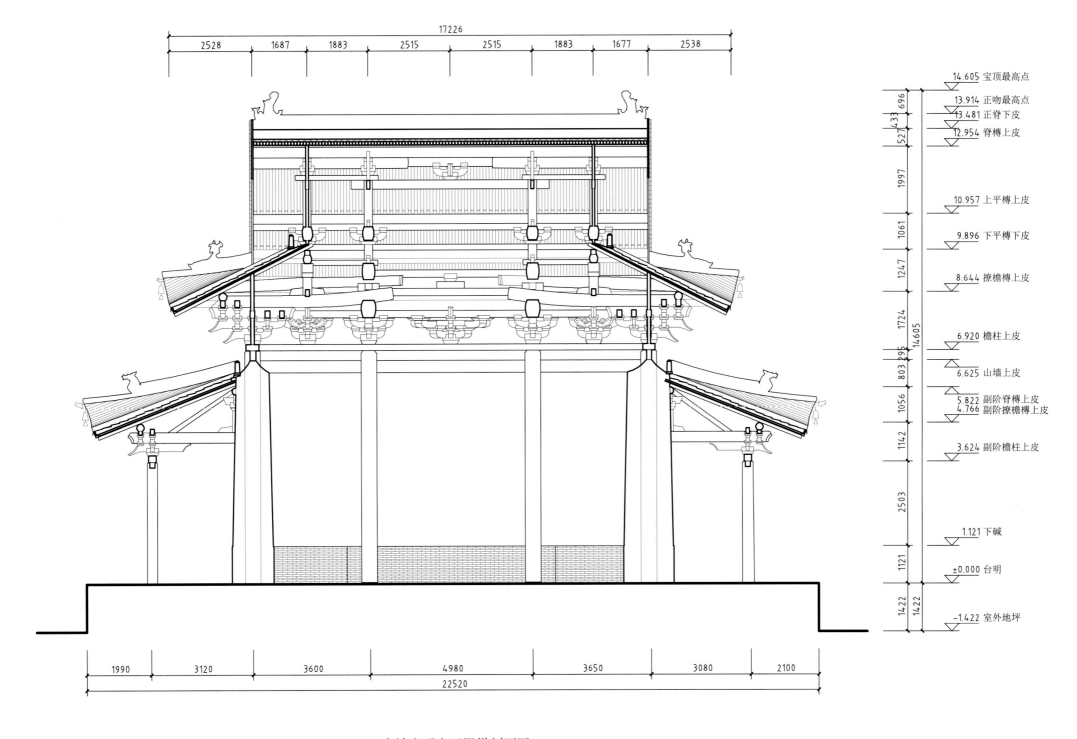

17226

| 2528 | 1687 | 1883 | 2515 | 2515 | 1883 | 1677 | 2538 |

14.605 宝顶最高点

13.914 正吻最高点

13.481 正脊下皮

12.954 脊槫上皮

10.957 上平槫上皮

9.896 下平槫下皮

8.644 撩檐槫上皮

6.920 檐柱上皮

6.625 山墙上皮

5.822 副阶脊槫上皮

4.766 副阶撩檐槫上皮

3.624 副阶檐柱上皮

1.121 下碱

±0.000 台明

-1.422 室外地坪

| 1990 | 3120 | 3600 | 4980 | 3650 | 3080 | 2100 |

22520

水神庙明应王殿纵剖面图
Longitudinal Section of the Mingying Wang Hall, Water God's Temple

0   1   3m

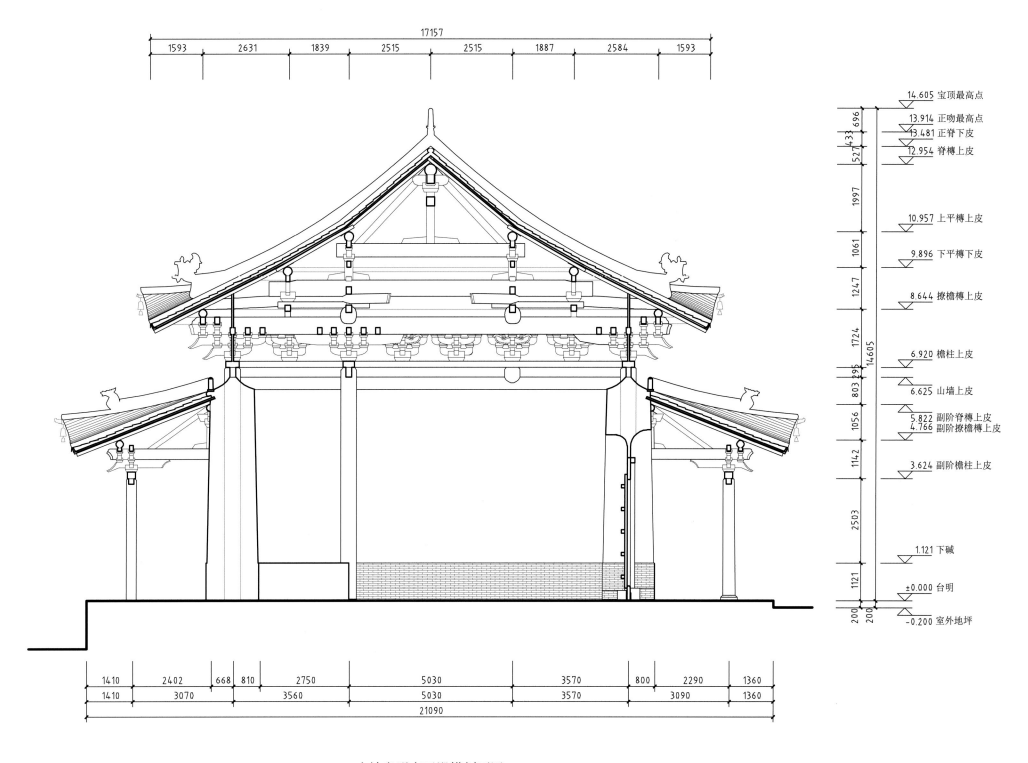

17157

1593　2631　1839　2515　2515　1887　2584　1593

14.605 宝顶最高点
13.914 正吻最高点
13.481 正脊下皮
12.954 脊槫上皮
10.957 上平槫上皮
9.896 下平槫下皮
8.644 撩檐槫上皮
6.920 檐柱上皮
6.625 山墙上皮
5.822 副阶脊槫上皮
4.766 副阶撩檐槫上皮
3.624 副阶檐柱上皮
1.121 下碱
±0.000 台明
-0.200 室外地坪

696　433　527　1997　1061　1247　1724　295　803　1056　1142　2503　1121　200　200

14605

1410　2402　668　810　2750　5030　3570　800　2290　1360
1410　3070　3560　5030　3570　3090　1360
21090

水神庙明应王殿横剖面图
Cross Section of the Mingying Wang Hall, Water God's Temple

0　1　3m

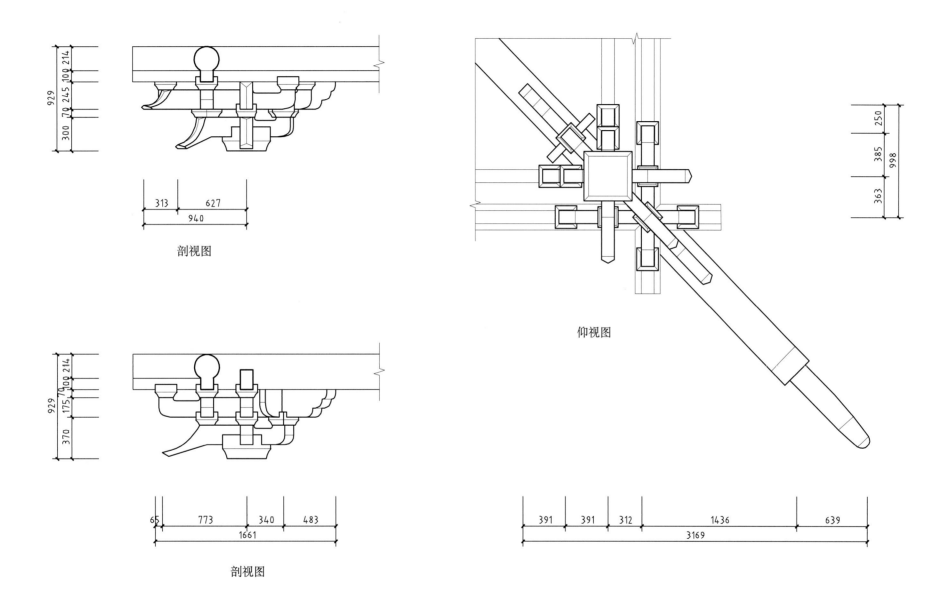

剖视图

剖视图

仰视图

水神庙明应王殿一层转角斗栱大样图

Detailed Drawing of the Lower-Eave Bracket Sets on Corners of the Mingying Wang Hall, Water God's Temple

0   0.2   1m

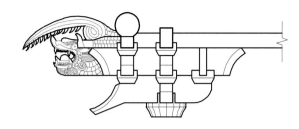

剖视图

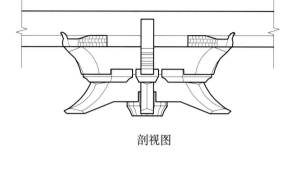

剖视图

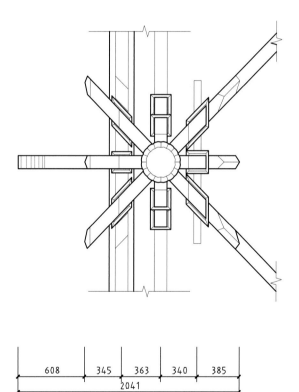

仰视图

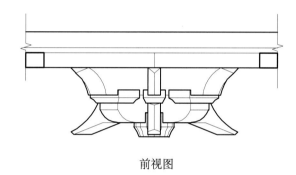

前视图

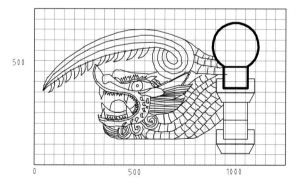

龙形耍头纹样

水神庙明应王殿一层前檐明间补间斗栱大样图

Detailed Drawing of the Lower-Eave Bracket Set in the Center of the Mingying Wang Hall's Front Elevation, Water God's Temple

0 0.2   1m

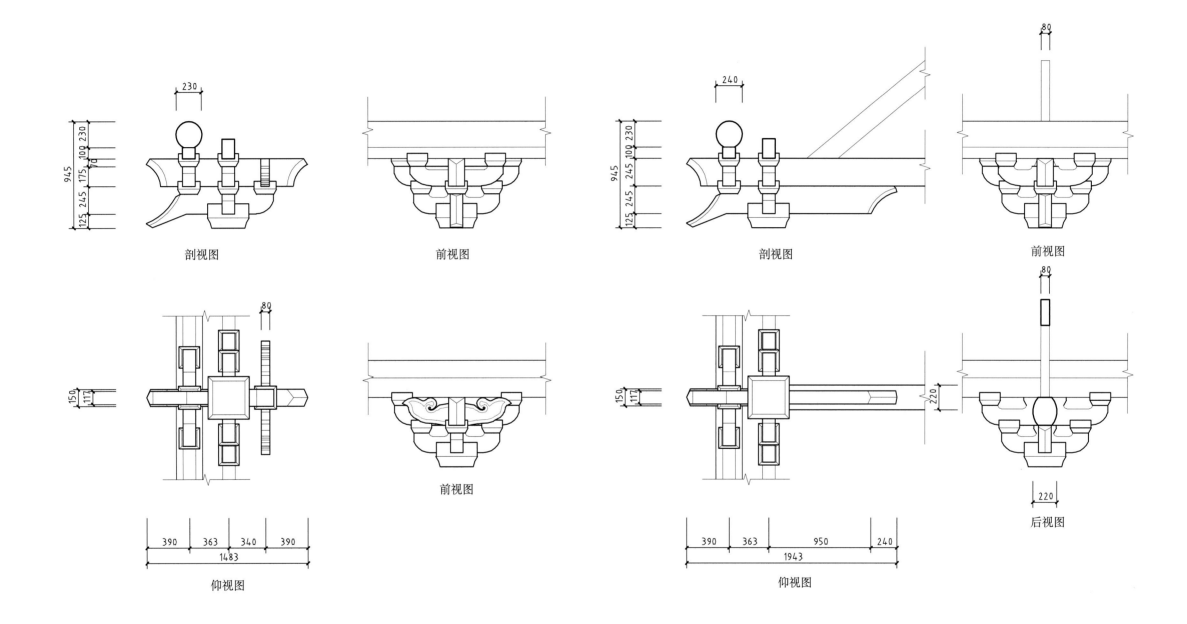

剖视图　　　　　前视图　　　　　剖视图　　　　　前视图

前视图　　　　　后视图

仰视图　　　　　仰视图

水神庙明应王殿一层山面明间补间斗栱大样图
Detailed Drawing of the Lower-Eave Bracket Sets in the Center of the Mingying Wang Hall's Gable Walls, Water God's Temple

水神庙明应王殿一层柱头斗栱大样图
Detailed Drawing of the Lower-Eave Bracket Sets on the Mingying Wang Hal's Columns, Water God's Temple

0　0.2　　　　　　1m

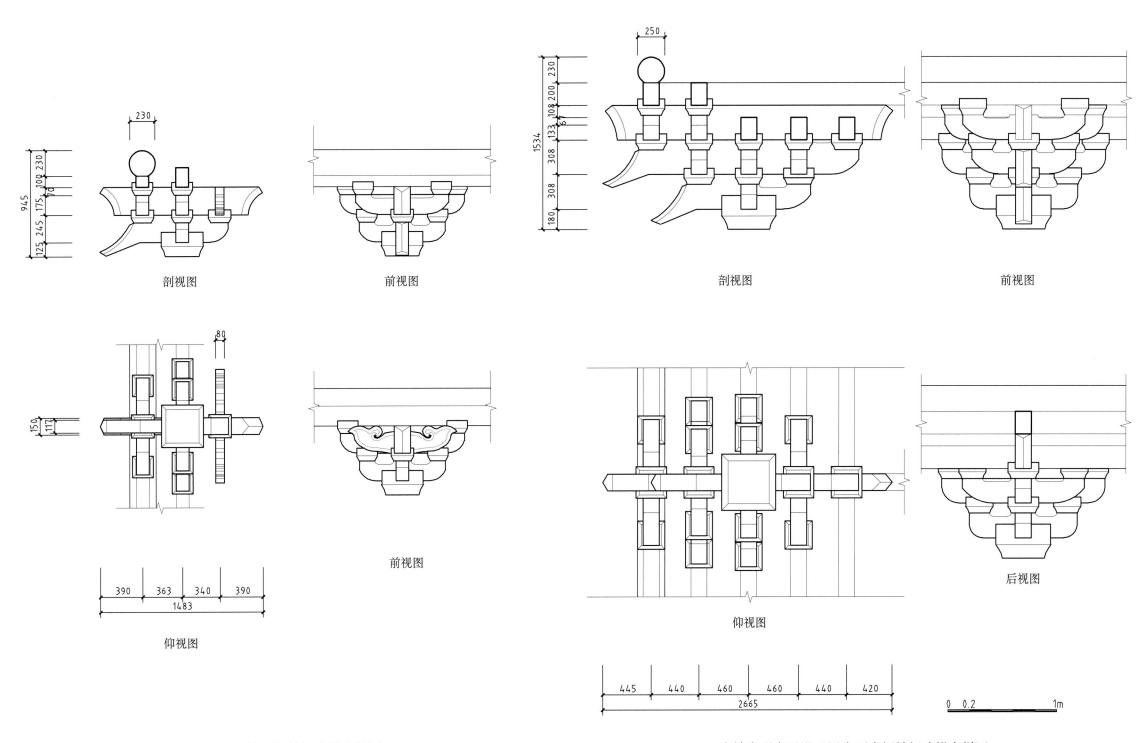

剖视图　　　　　　　前视图　　　　　　　剖视图　　　　　　　前视图

前视图　　　　　　　仰视图

仰视图　　　　　　　后视图

水神庙明应王殿一层后檐明间补间斗栱大样图
Detailed Drawing of the Lower-Eave Bracket Set in the Center of the Mingying Wang Hall's
Back Elevation, Water God's Temple

水神庙明应王殿二层山面次间补间斗栱大样图
Detailed Drawing of the Upper-Eave Bracket Sets between Columns of the Bays next to the Central Bay of the Mingying
Wang Hall's Gable Walls, Water God's Temple

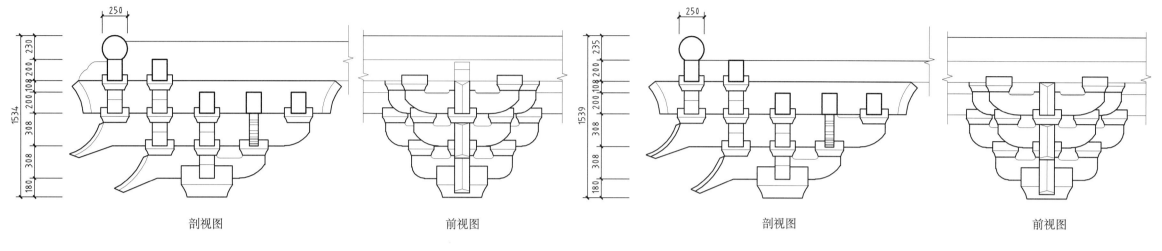

剖视图　　　　　　　　前视图　　　　　　　　　剖视图　　　　　　　　前视图

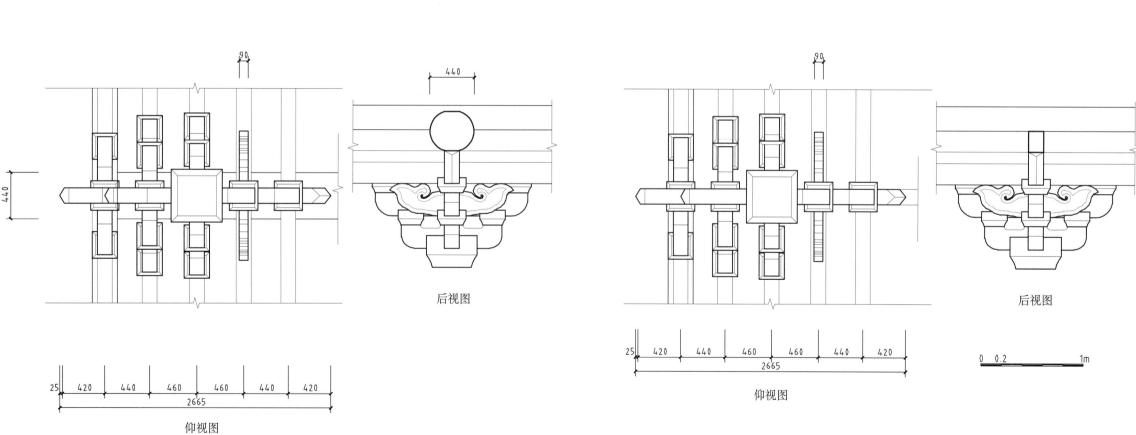

仰视图　　　　　　　　　　　后视图　　　　　　　　　　　仰视图　　　　　　　　　　后视图

水神庙明应王殿二层山面明间柱头斗栱大样图（一）
Detailed Drawing of the Upper-Eave Bracket Sets on the Central Bays' Columns of the Mingying Wang Hall's Gable Walls, Water God's Temple (1)

水神庙明应王殿二层山面明间补间斗栱大样图
Detailed Drawing of the Upper-Eave Bracket Sets between the Central Bays' Columns of the Mingying Wang Hall's Gable Walls, Water God's Temple

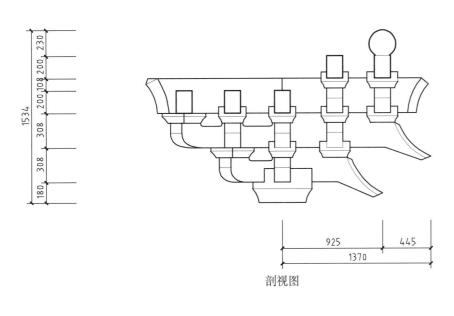

剖视图

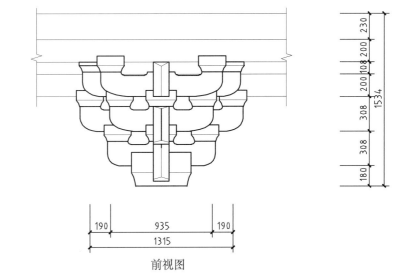

前视图

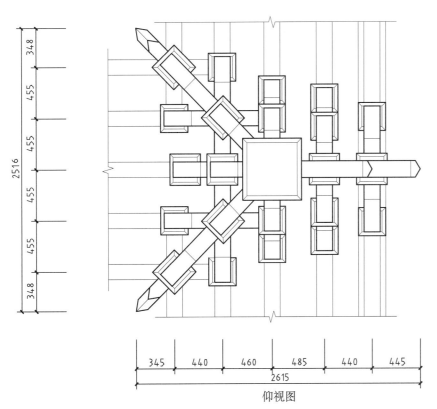

仰视图

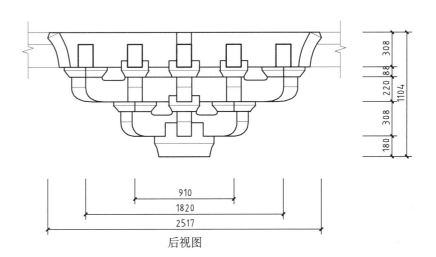

后视图

水神庙明应王殿二层山面明间柱头斗栱大样图（二）

Detailed Drawing of the Upper-Eave Bracket Sets on the Central Bays' Columns of the Mingying Wang Hall's Gable
Walls, Water God's Temple (2)

0   0.2   1m

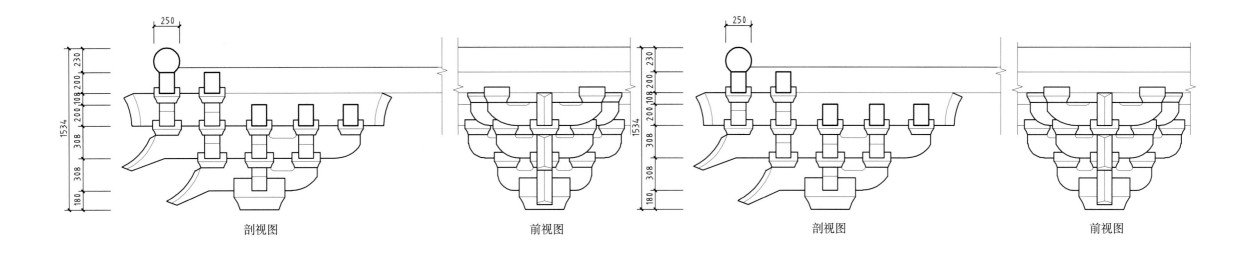

剖视图　　　　　　　前视图　　　　　　　剖视图　　　　　　　前视图

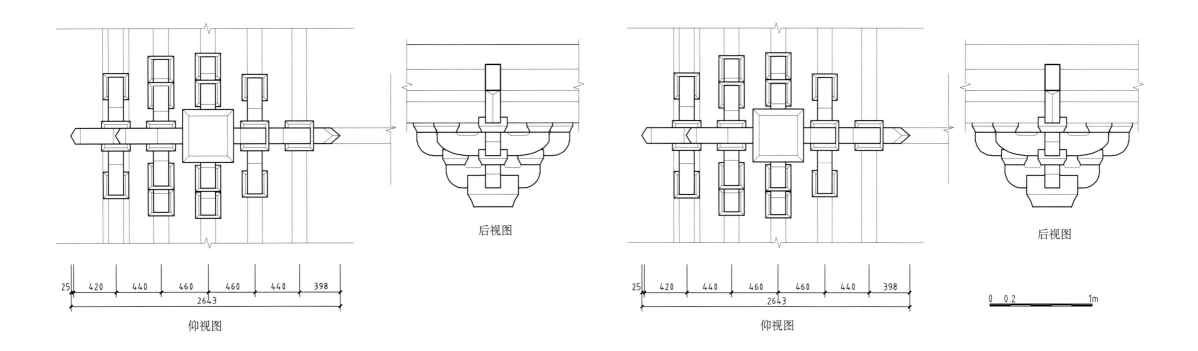

后视图　　　　　　　　　　　　　　后视图

仰视图　　　　　　　　　　　　　仰视图

0　0.2　　　1m

水神庙明应王殿二层前檐次间补间斗栱大样图
Detailed Drawing of the Upper-Eave Bracket Sets between Columns of the Bays next to the Central Bay of the
Mingying Wang Hall's Front Elevation, Water God's Temple

水神庙明应王殿二层后檐次间补间斗栱大样图
Detailed Drawing of the Upper-Eave Bracket Sets between Columns of the Bays next to the Central Bay of the
Mingying Wang Hall's Back Elevation, Water God's Temple

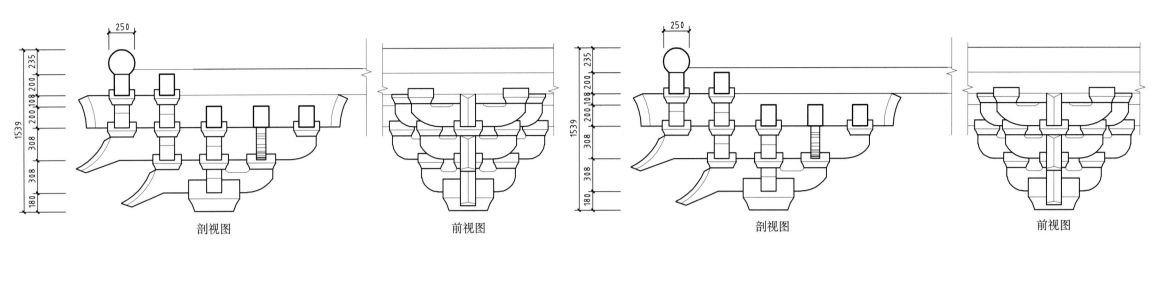

剖视图   前视图   剖视图   前视图

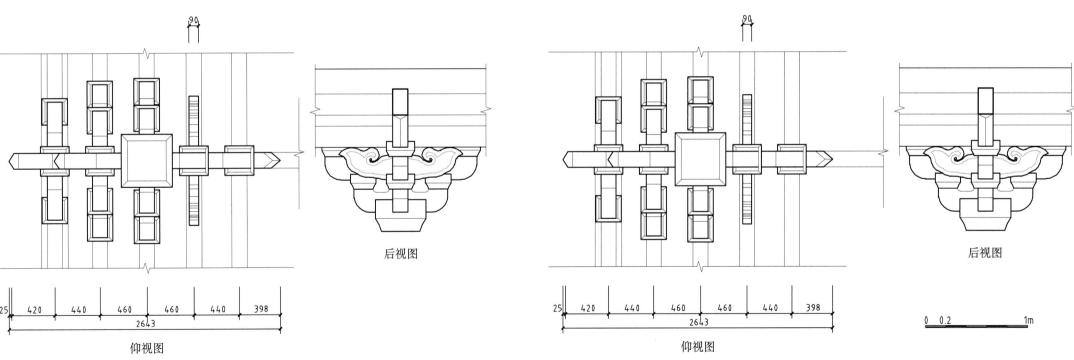

后视图   后视图

仰视图   仰视图

水神庙明应王殿二层前檐明间补间斗栱大样图
Detailed Drawing of the Upper-Eave Bracket Sets between the Central Bays' Columns of the
Mingying Wang Hall's Front Elevation, Water God's Temple

水神庙明应王殿二层后檐明间补间斗栱大样图
Detailed Drawing of the Upper-Eave Bracket Sets between the Central Bays' Columns of the Mingying Wang Hall's
Back Elevation, Water God's Temple

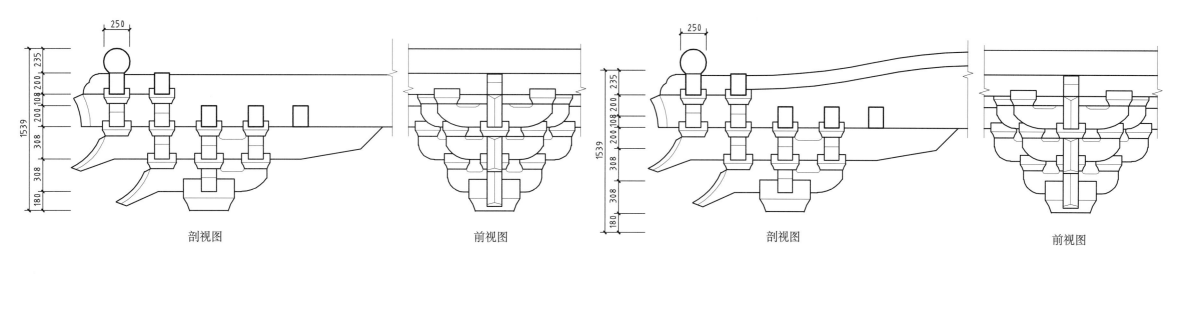

剖视图   前视图   剖视图   前视图

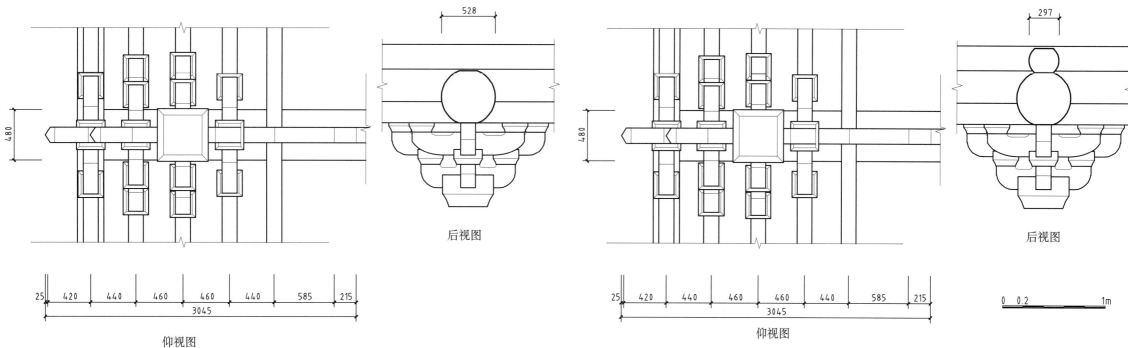

后视图   后视图

0   0.2   1m

仰视图   仰视图

水神庙明应王殿二层前檐明间柱头斗栱大样图

Detailed Drawing of the Upper-Eave Bracket Sets on the Central Bays' Columns of the Mingying Wang Hall's Front Elevation, Water God's Temple

水神庙明应王殿二层后檐明间柱头斗栱大样图

Detailed Drawing of the Upper-Eave Bracket Sets on the Central Bays' Columns of the Mingying Wang Hall's Back Elevation, Water God's Temple

1104
180
308
308
200
108
剖视图
前视图
1147
仰视图
420
440
460
460
440
420
2640

水神庙明应王殿二层后檐金柱次间补间斗栱大样图
Detailed Drawing of the Upper-Eave Bracket Sets between Hypostyle Columns of the Bays next to the Central Bay of the Mingying Wang Hall's Back Elevation, Water God's Temple

1141
180
220
88
220
88
200
145
剖视图
后视图
仰视图
805
440
460
460
440
805
3410
0 0.2 1m

水神庙明应王殿二层后檐金柱柱头斗栱大样图
Detailed Drawing of the Upper-Eave Bracket Sets on Hypostyle Columns of the Bays next to the Central Bay of the Mingying Wang Hall's Back Elevation, Water God's Temple

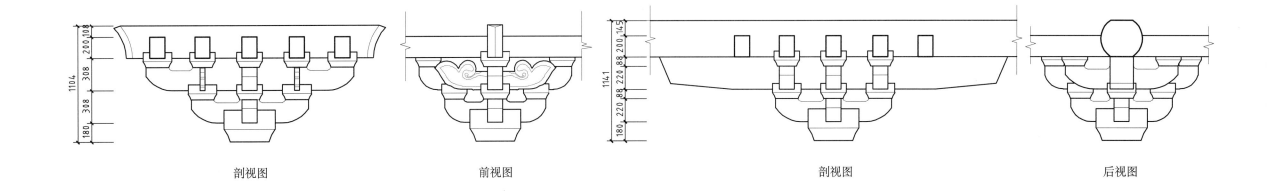

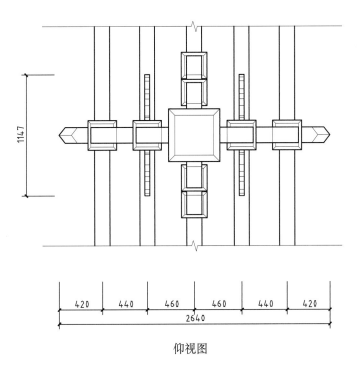

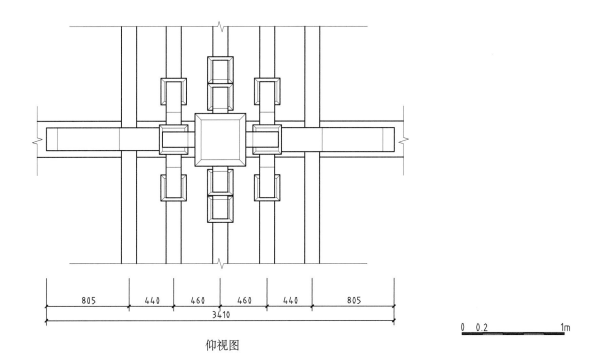

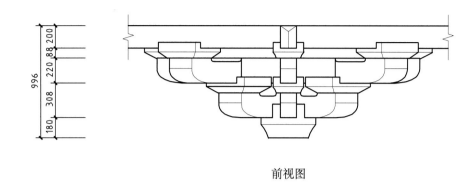

前视图

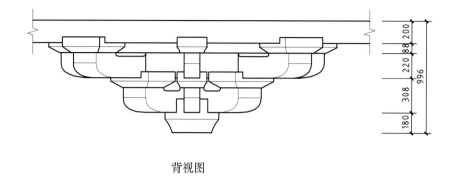

背视图

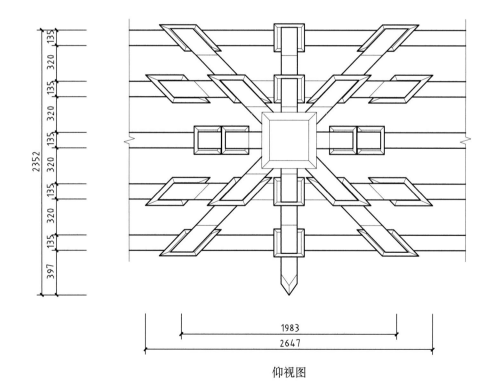

仰视图

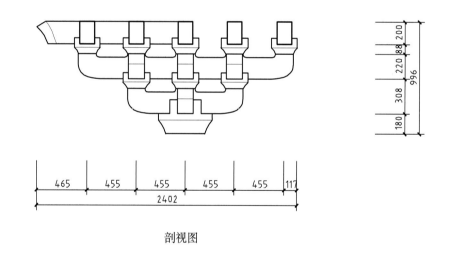

剖视图

水神庙明应王殿金柱明间补间斗栱大样图

Detailed Drawing of the Bracket Sets between Central Bay's Hypostyle Columns of the Mingying Wang Hall, Water God's Temple

0   0.2   1m

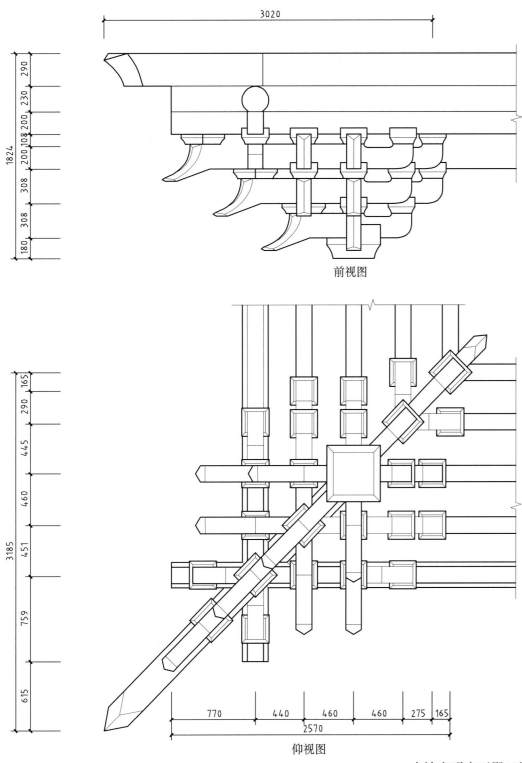

前视图

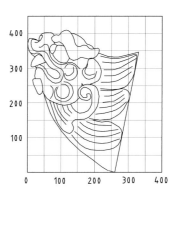

背视图

仰视图

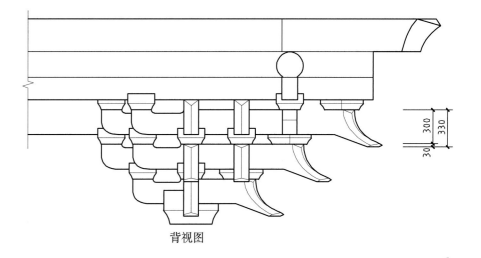

水神庙明应王殿二层转角柱头斗栱大样图
Detailed Drawing of the Upper-Eave Bracket Sets on Corner Columns of the Mingying Wang Hall, Water God's Temple

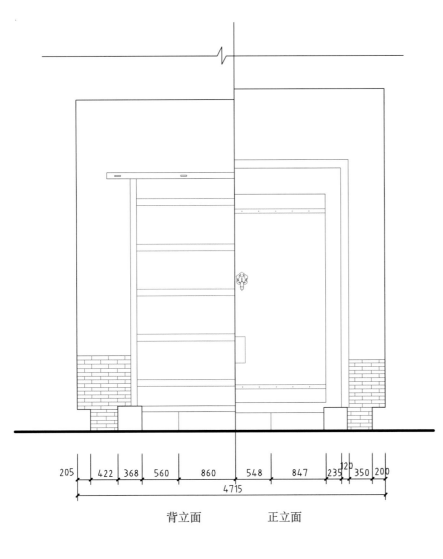

背立面　　　正立面

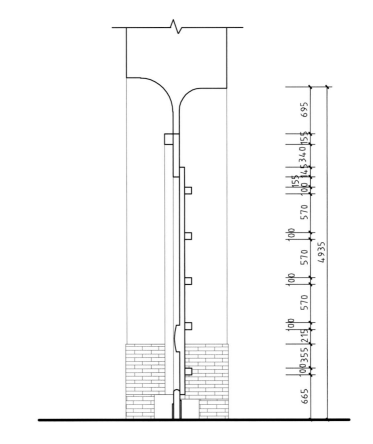

1—1 剖面

门坎石刻

门枕石纹样

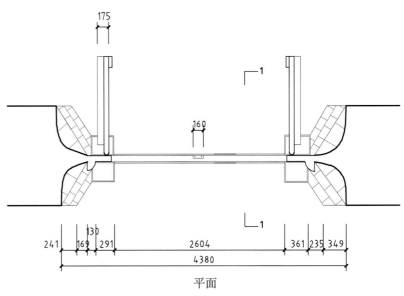

平面

水神庙明应王殿大门大样图
Detailed Drawing of the Gate of the Mingying Wang Hall, Water God's Temple

0　　0.5　　1.5m

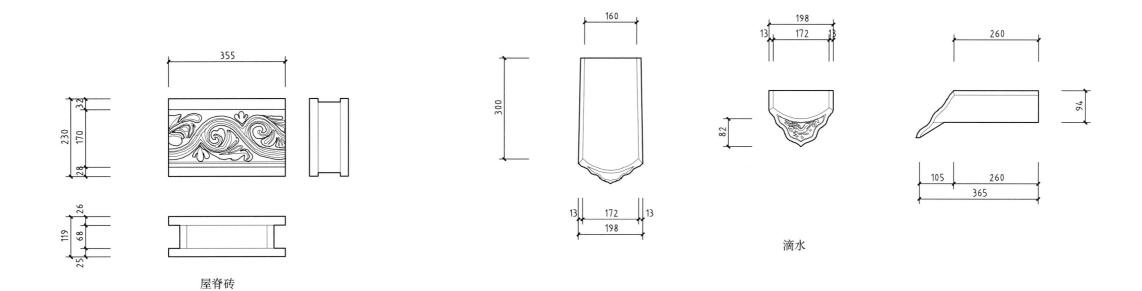

屋脊砖

滴水

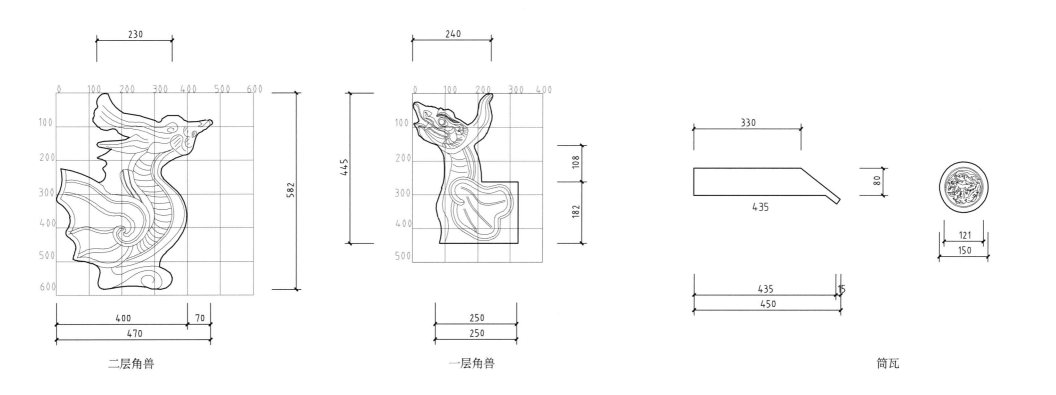

二层角兽

一层角兽

筒瓦

水神庙明应王殿细部大样图
Detailed Drawing of the Mingying Wang Hall, Water God's Temple

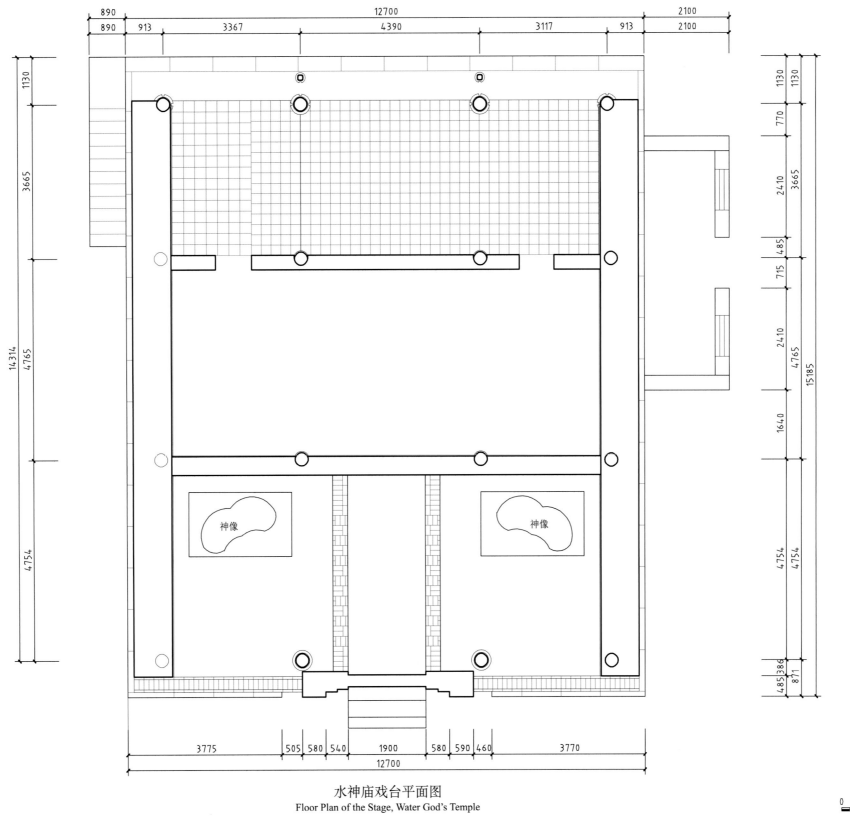

水神庙戏台平面图
Floor Plan of the Stage, Water God's Temple

神像

神像

N

0 1 2m

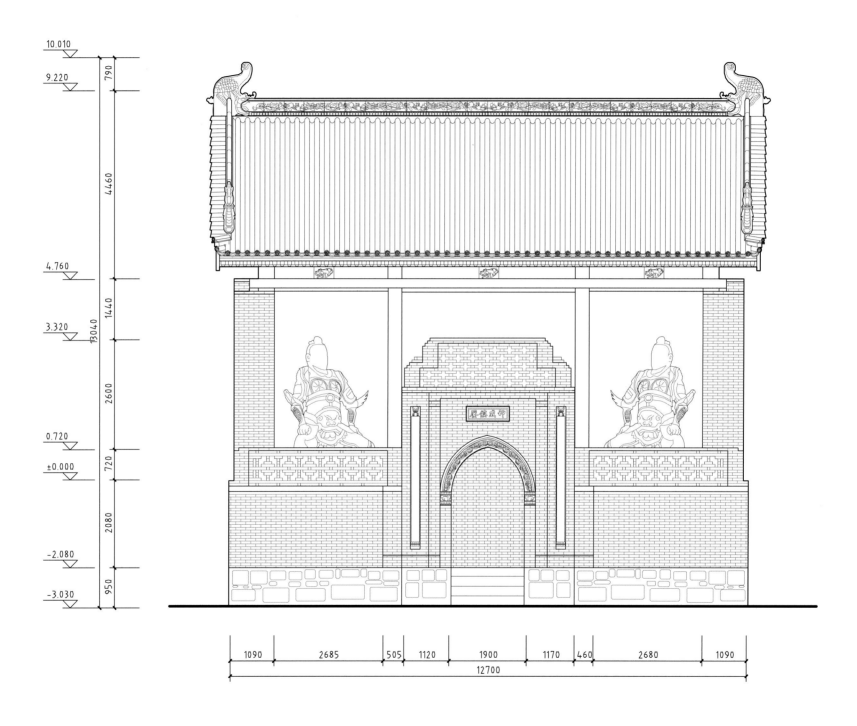

水神庙戏台南立面图
Southern Elevation of the Stage, Water God's Temple

0 1 2m

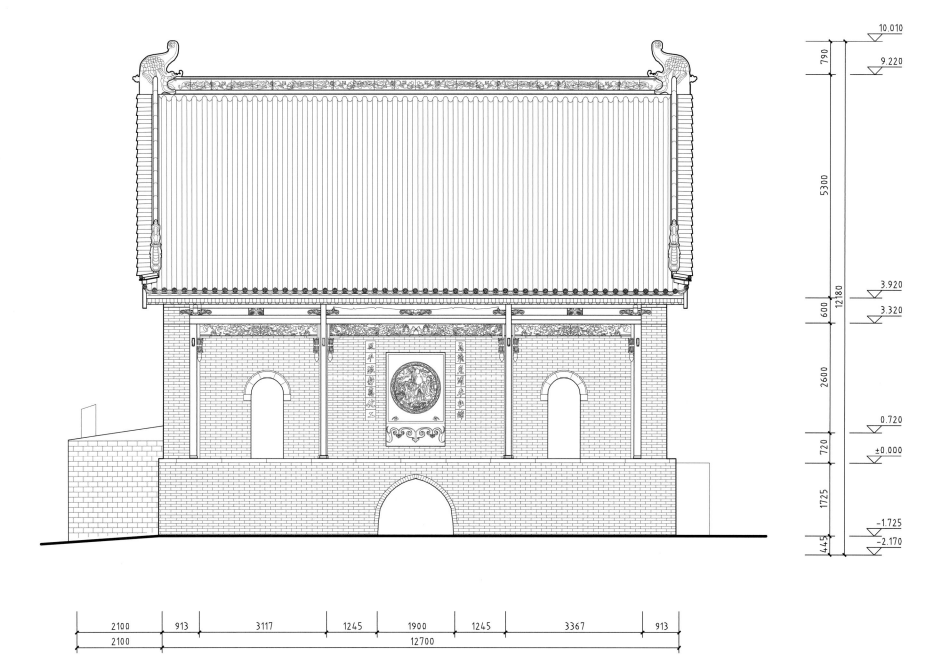

10.010

790

9.220

5300

12180

3.920

600

3.320

2600

0.720

720

±0.000

1725

-1.725

445

-2.170

| 2100 | 913 | 3117 | 1245 | 1900 | 1245 | 3367 | 913 |

2100 | 12700

水神庙戏台北立面图
Northern Elevation of the Stage, Water God's Temple

0   1   2m

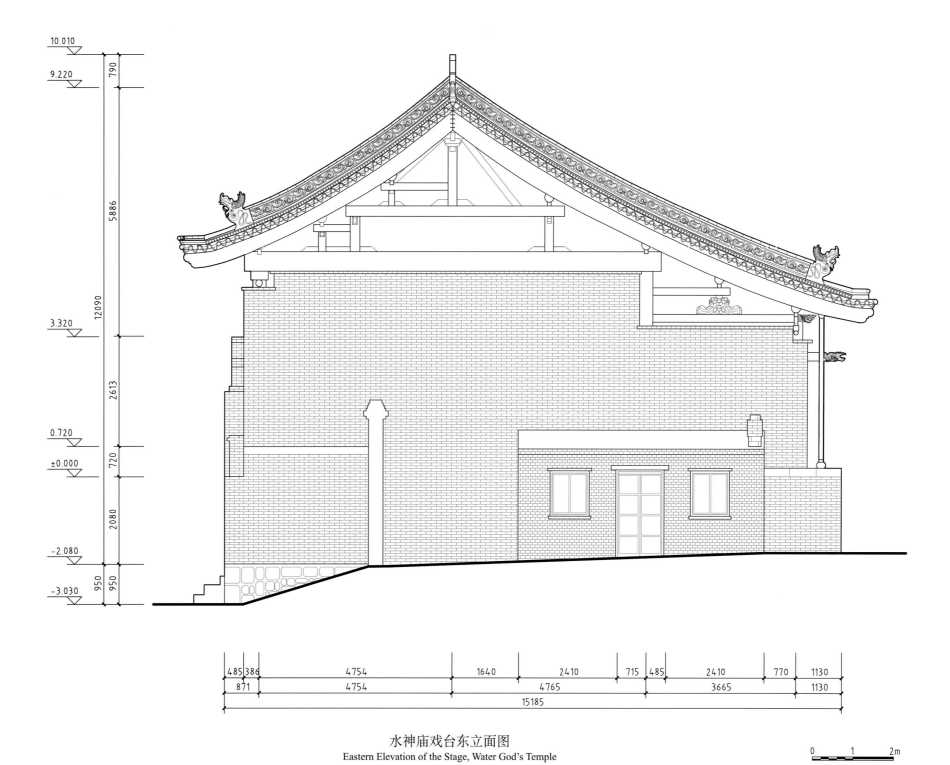

10.010

9.220

790

5886

12090

3.320

2613

0.720

±0.000

720

-2.080

2080

-3.030

950

950

485 386
871

4754
4754

1640

2410
4765

715 485

2410
3665

770

1130
1130

15185

水神庙戏台东立面图
Eastern Elevation of the Stage, Water God's Temple

0   1   2m

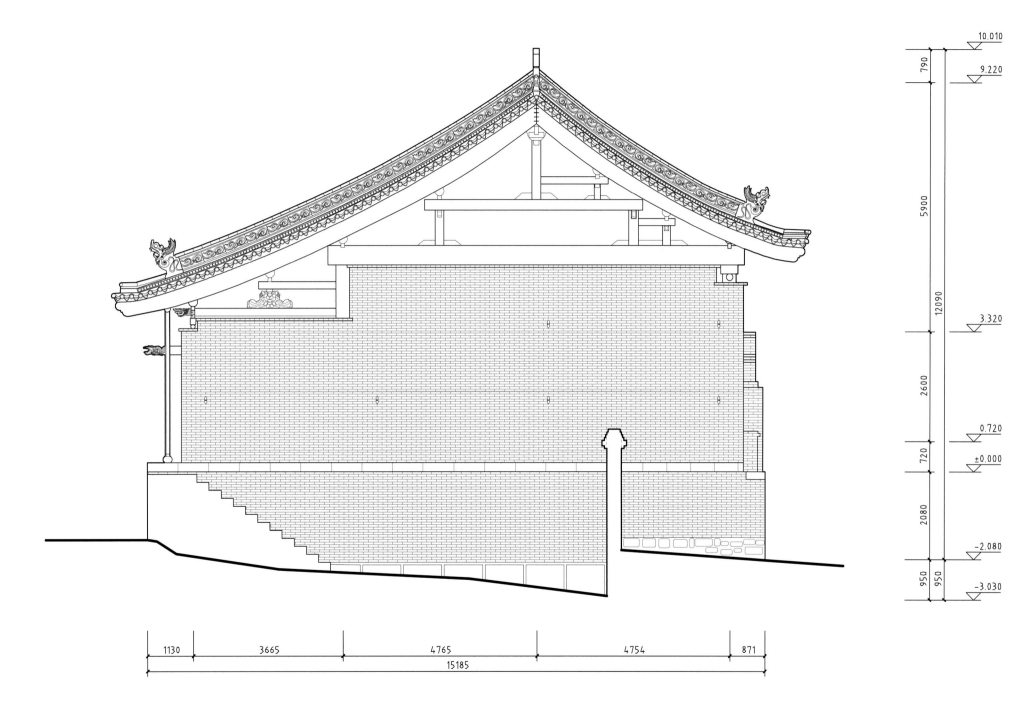

水神庙戏台西立面图
Western Elevation of the Stage, Water God's Temple

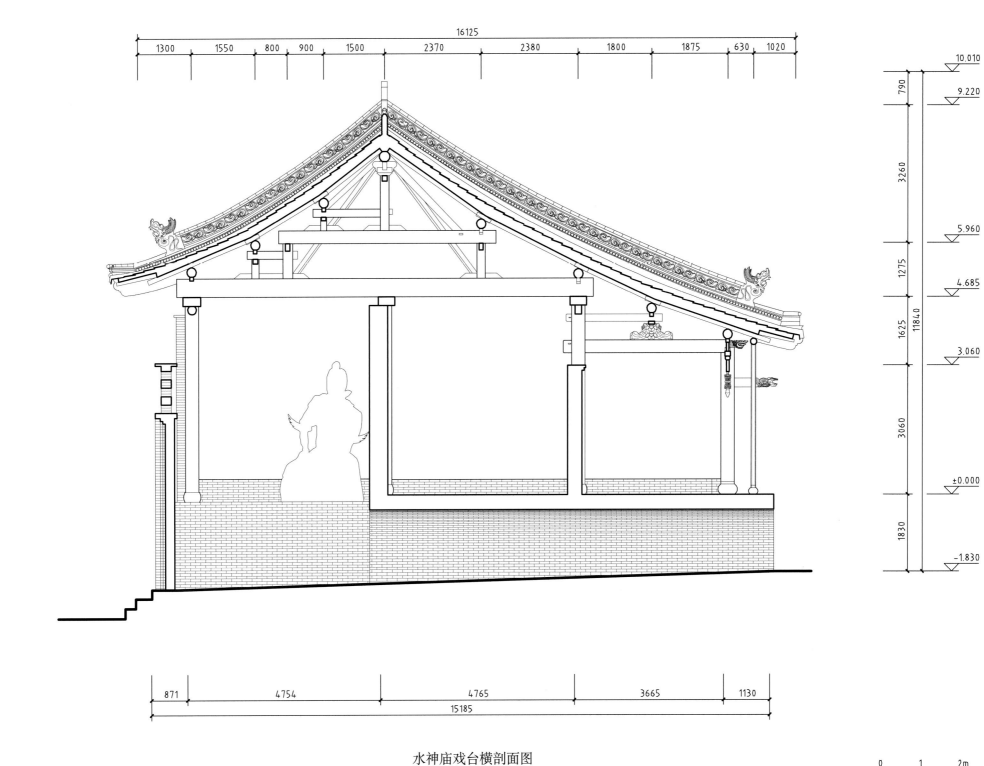

水神庙戏台横剖面图
Cross Section of the Stage, Water God's Temple

0 1 2m

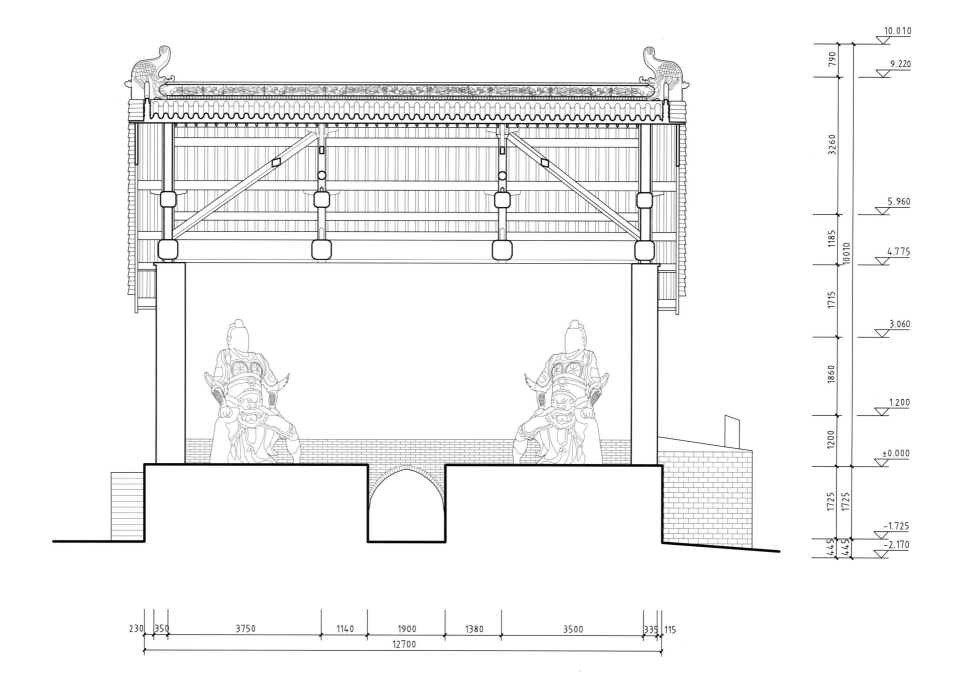

水神庙戏台纵剖面图
Longitudinal Section of the Stage, Water God's Temple

0   1   2m

广胜下寺

Lower Guangsheng
Monastery

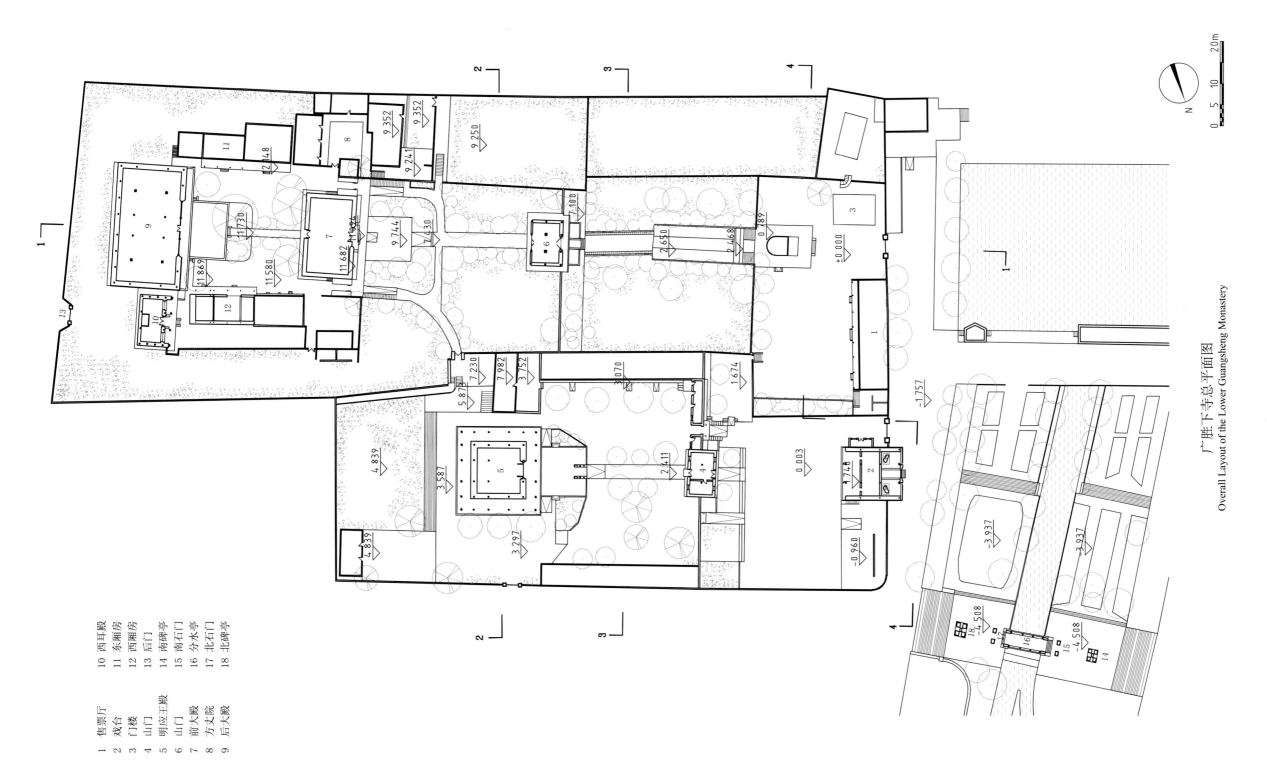

1　售票厅　　10　西耳殿
2　戏台　　　11　东厢房
3　门楼　　　12　西厢房
4　山门　　　13　后门
5　明应王殿　14　南碑亭
6　山门　　　15　南石门
7　前大殿　　16　分水亭
8　方丈院　　17　北石门
9　后大殿　　18　北碑亭

广胜下寺总平面图
Overall Layout of the Lower Guangsheng Monastery

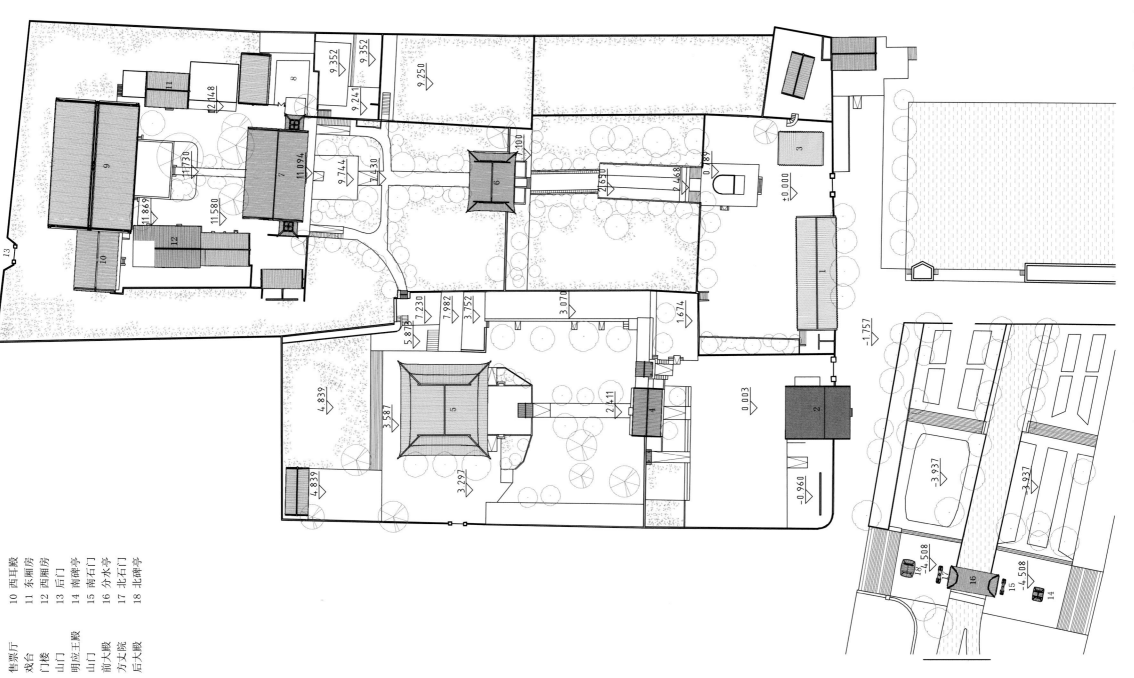

1	售票厅	10	西耳殿房
2	戏台	11	东厢房
3	门楼	12	西厢房
4	山门	13	后门
5	明应王殿	14	南碑亭
6	山门	15	南右门
7	前大殿	16	分水亭
8	方丈院	17	北右门
9	后大殿	18	北碑亭

广胜下寺屋顶总平面图

Overall Topview of Roofs, Lower Guangsheng Monastery

7.430

9.744

11.730

12.427

0 1 5m

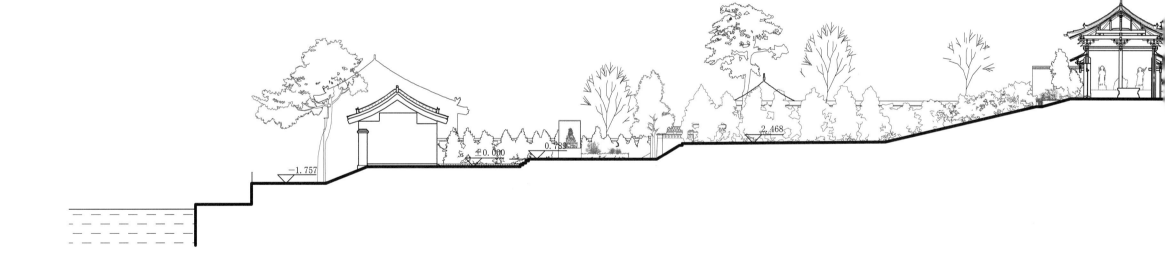

-1.757

±0.000

0.789

2.468

广胜下寺院落总纵剖面图（1-1）
1-1 Overall Longitudinal Section, Lower Guangsheng Monastery

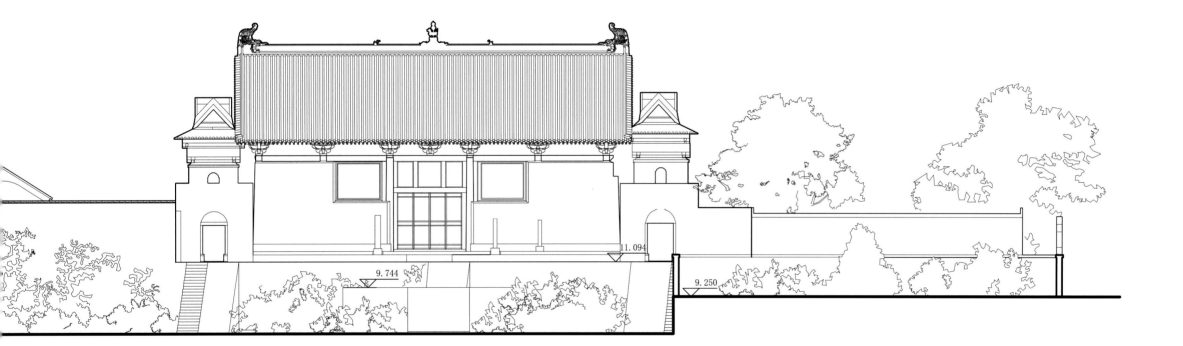

11.094

9.744

9.250

0 1   5m

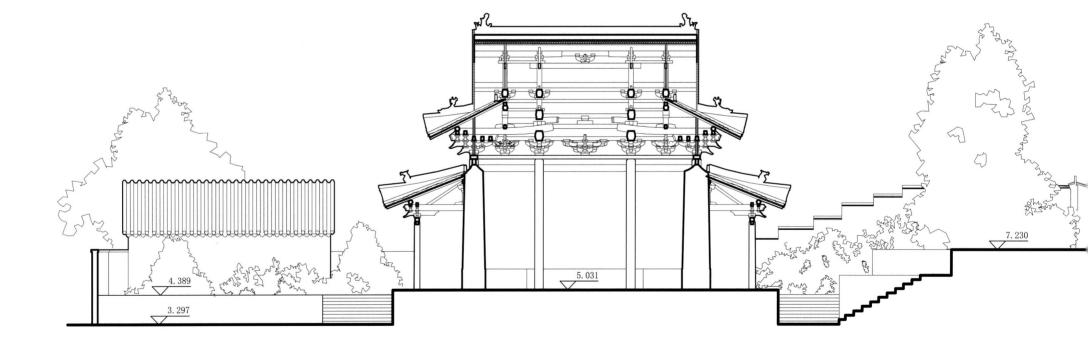

4.389

3.297

5.031

7.230

广胜下寺院落北段总横剖面图（2—2）
2-2 Overall Cross Section of the Northern Part, Lower Guangsheng Monastery

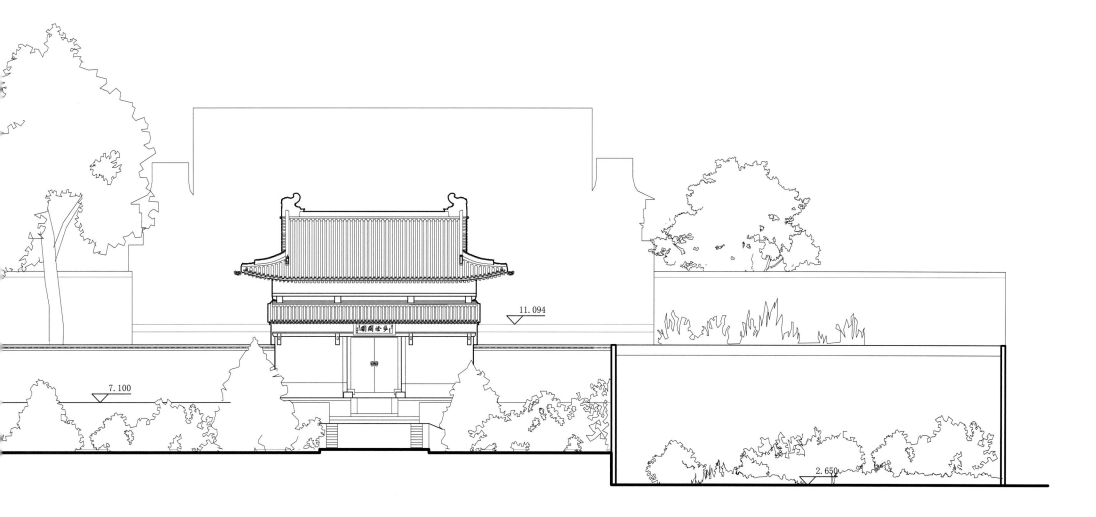

11.094

7.100

2.650

0 1   5m

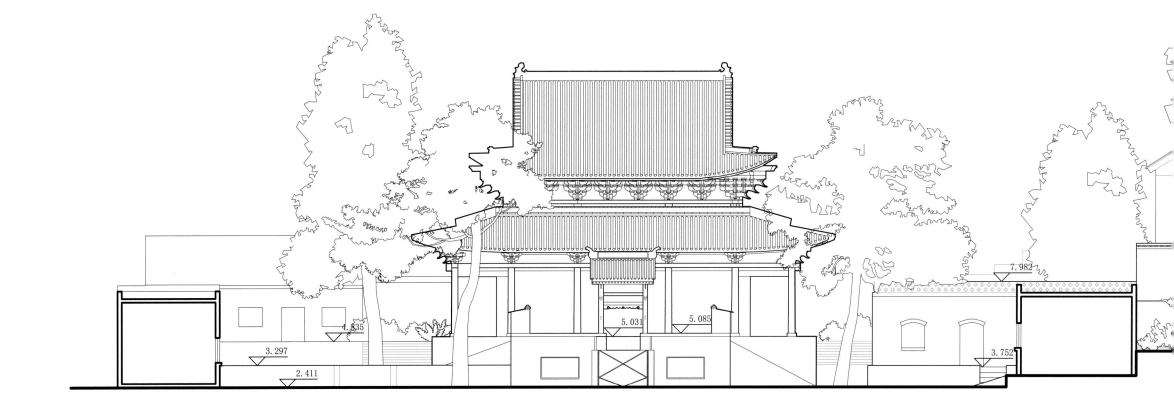

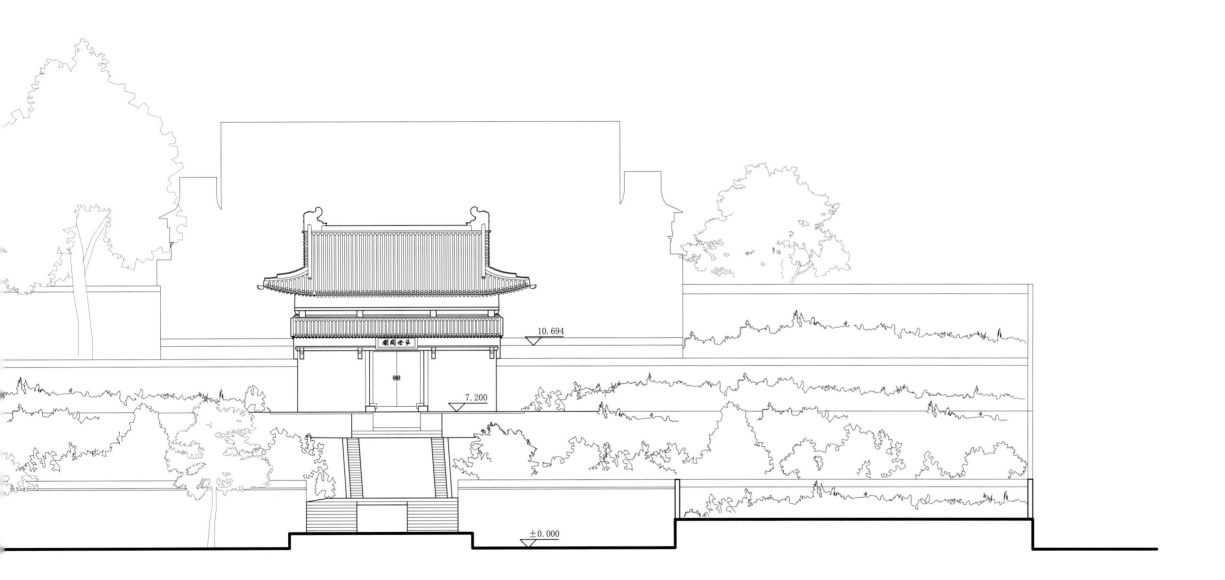

10.694

7.200

±0.000

0 1   5m

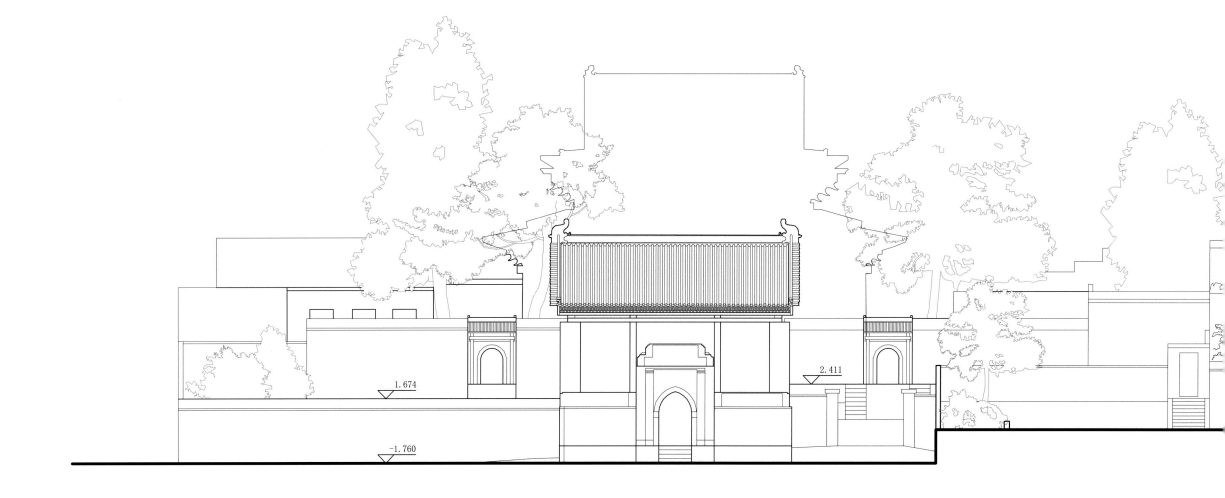

1.674

2.411

-1.760

广胜下寺院落南段总横剖面图（4-4）

4-4 Overall Cross Section of the Southern Part, Lower Guangsheng Monastery

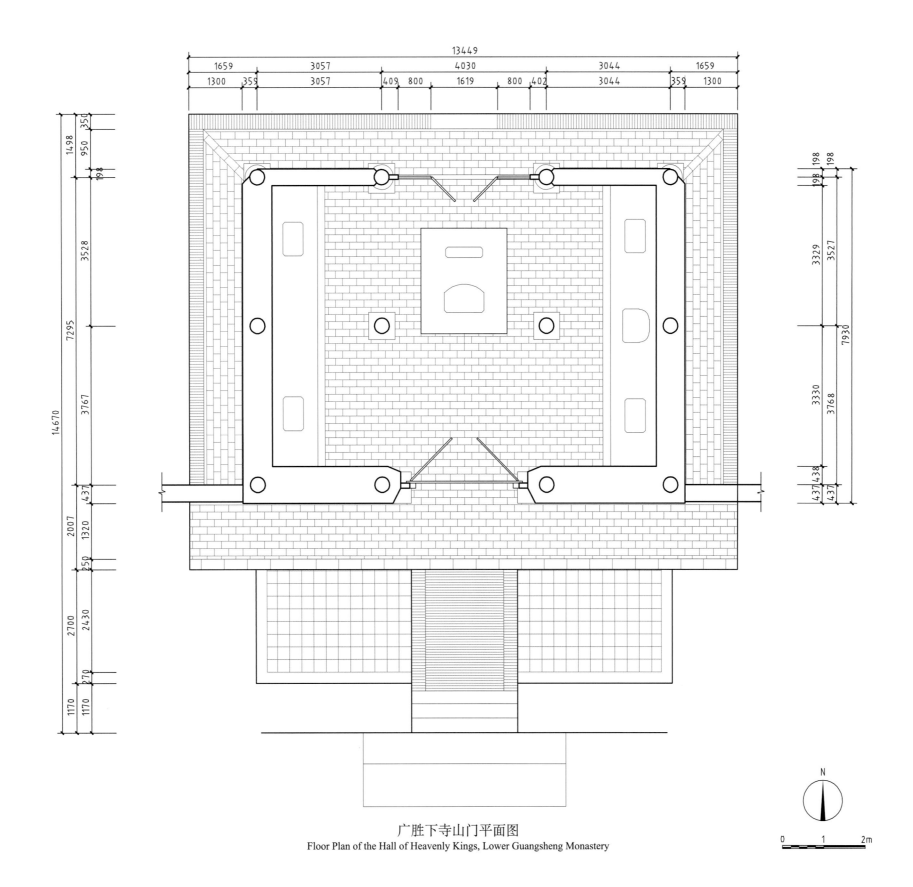

广胜下寺山门平面图
Floor Plan of the Hall of Heavenly Kings, Lower Guangsheng Monastery

0 1 2m

N

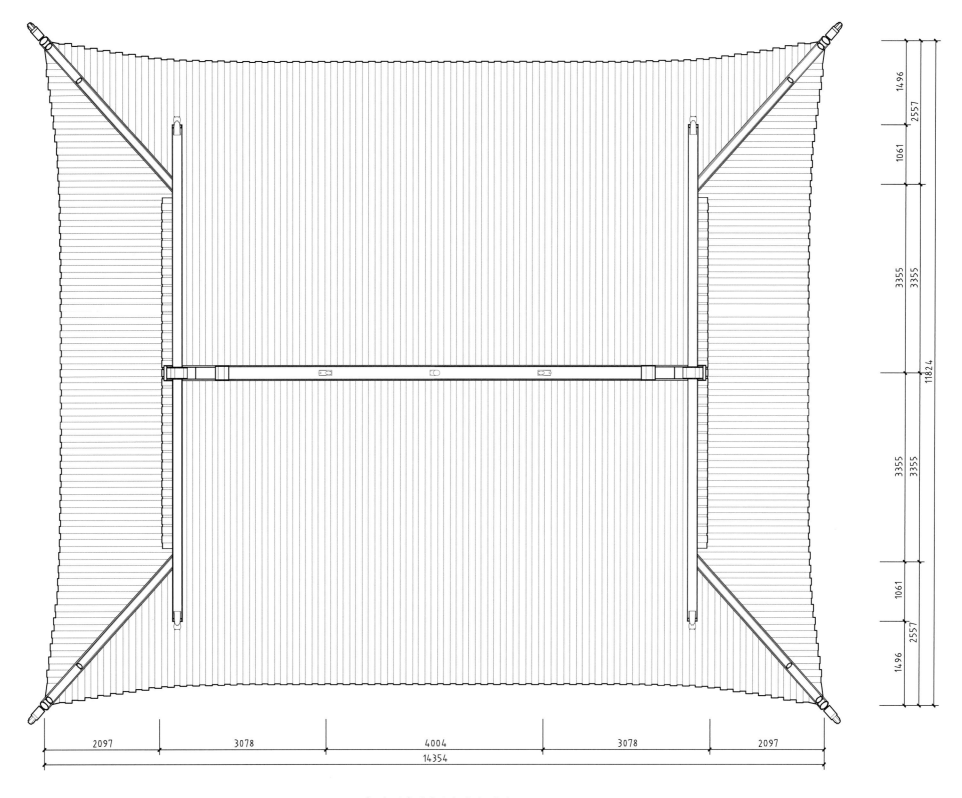

广胜下寺山门屋顶平面图
Top View of the Roof of the Hall of Heavenly Kings, Lower Guangsheng Monastery

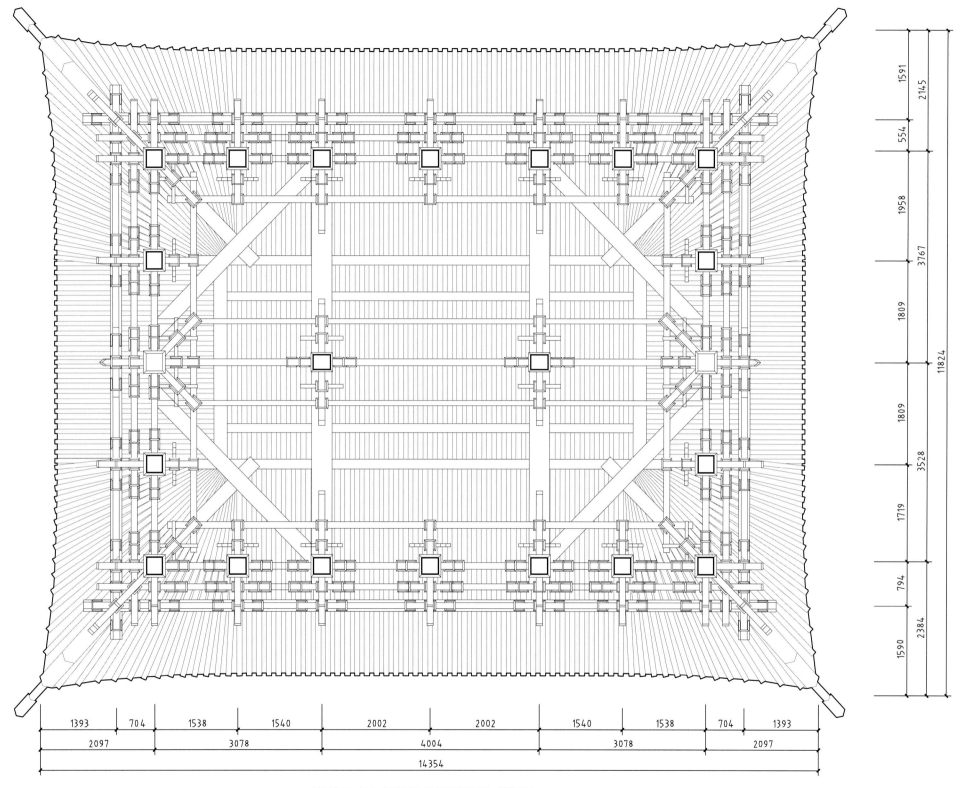

1591
2145
554
1958
3767
1809
11824
1809
3528
1719
794
2384
1590

1393 704 1538 1540 2002 2002 1540 1538 704 1393
2097 3078 4004 3078 2097
14354

广胜下寺山门梁架仰视平面图（镜像）
Bottom View of the Truss of the Hall of Heavenly Kings, Lower Guangsheng Monastery (Mirror Image)

0 0.5 1 2m

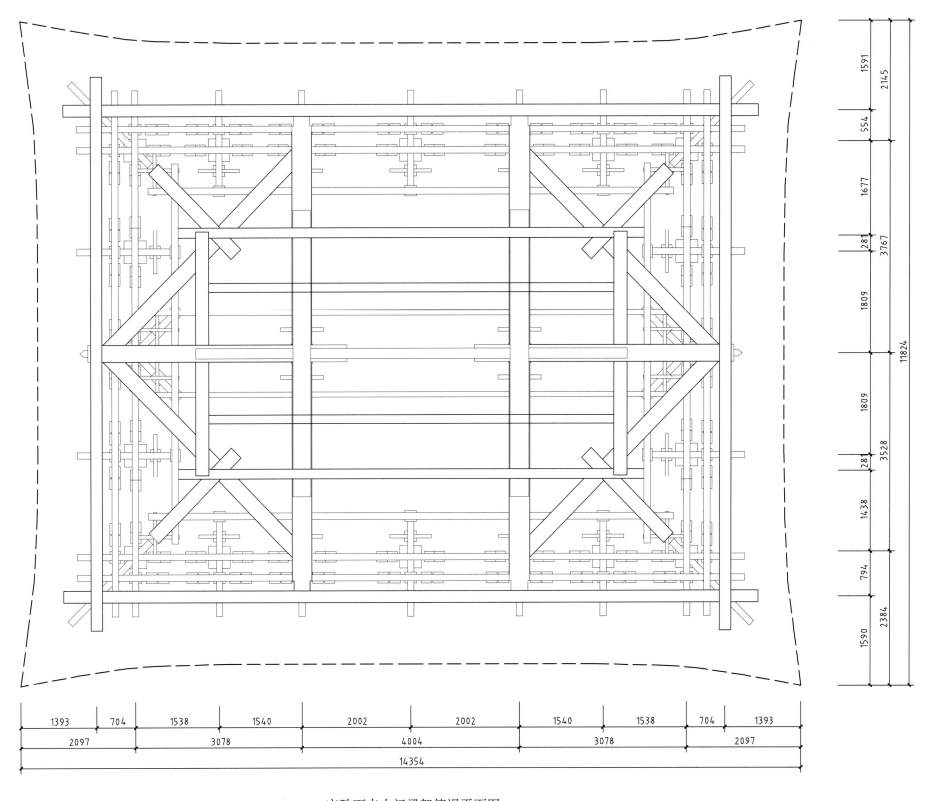

广胜下寺山门梁架俯视平面图
Top View of the Truss of the Hall of Heavenly Kings, Lower Guangsheng Monastery

0 0.5 1 2m

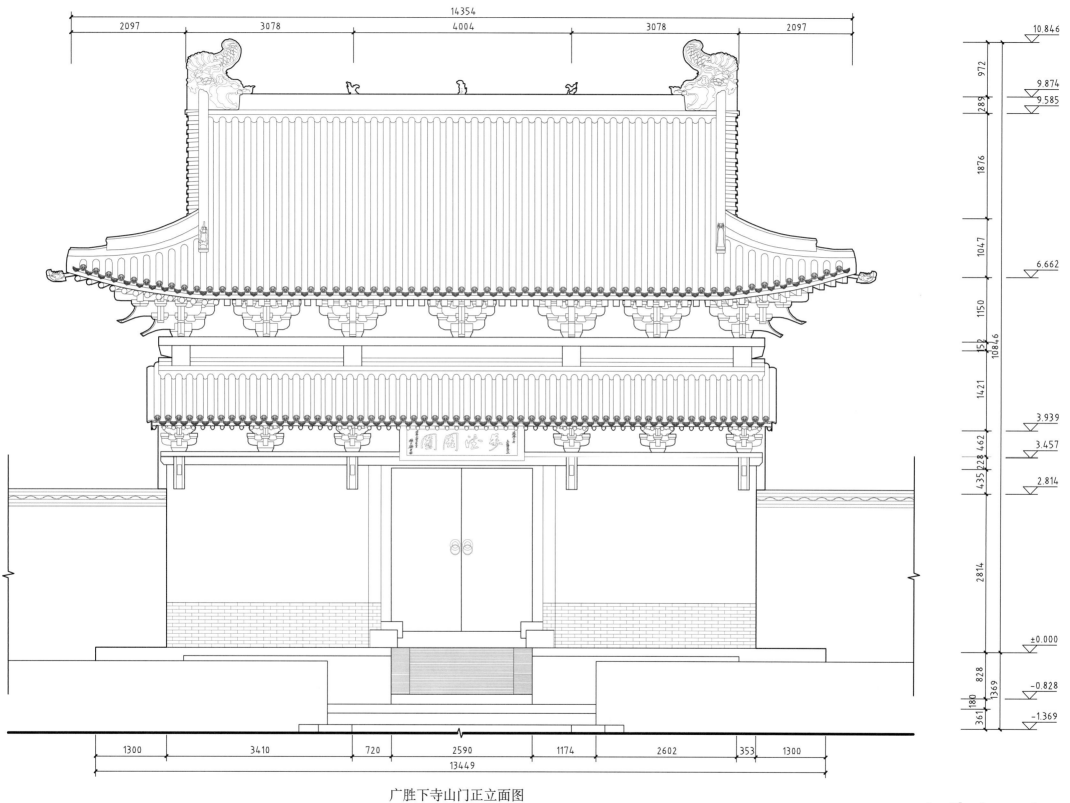

广胜下寺山门正立面图
Front Elevation of the Hall of Heavenly Kings, Lower Guangsheng Monastery

0 0.5 1 2m

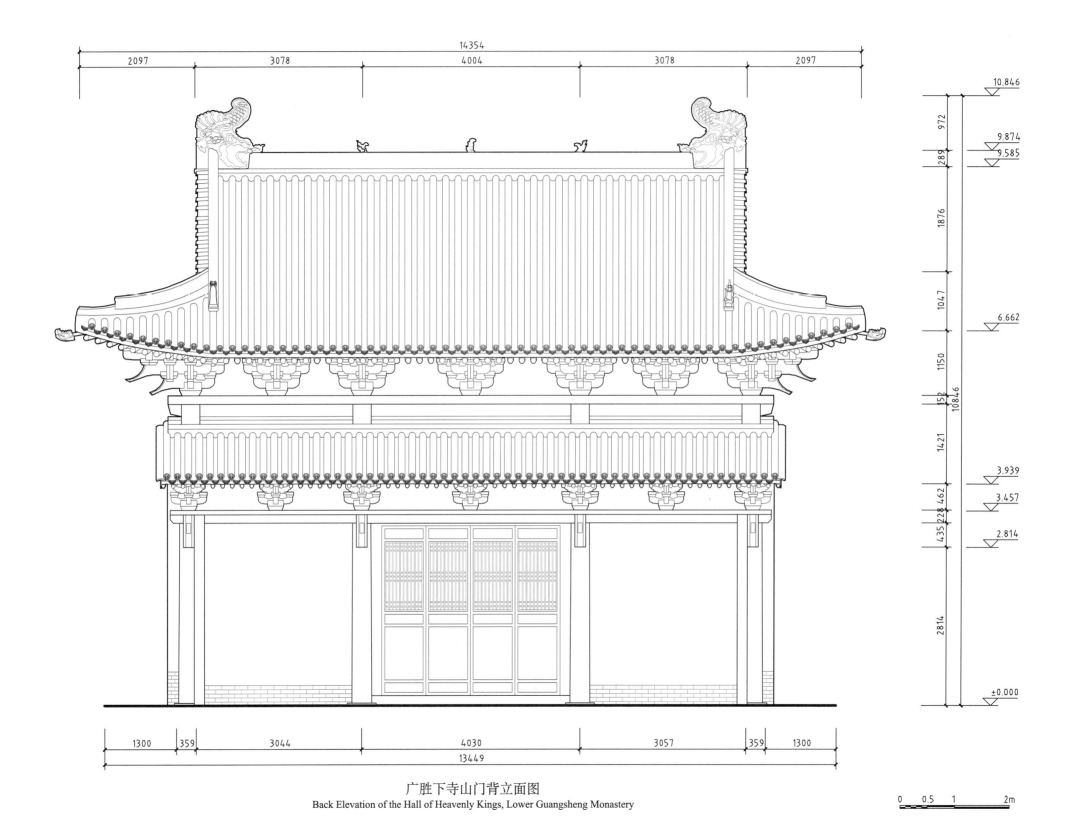

广胜下寺山门背立面图
Back Elevation of the Hall of Heavenly Kings, Lower Guangsheng Monastery

0 0.5 1 2m

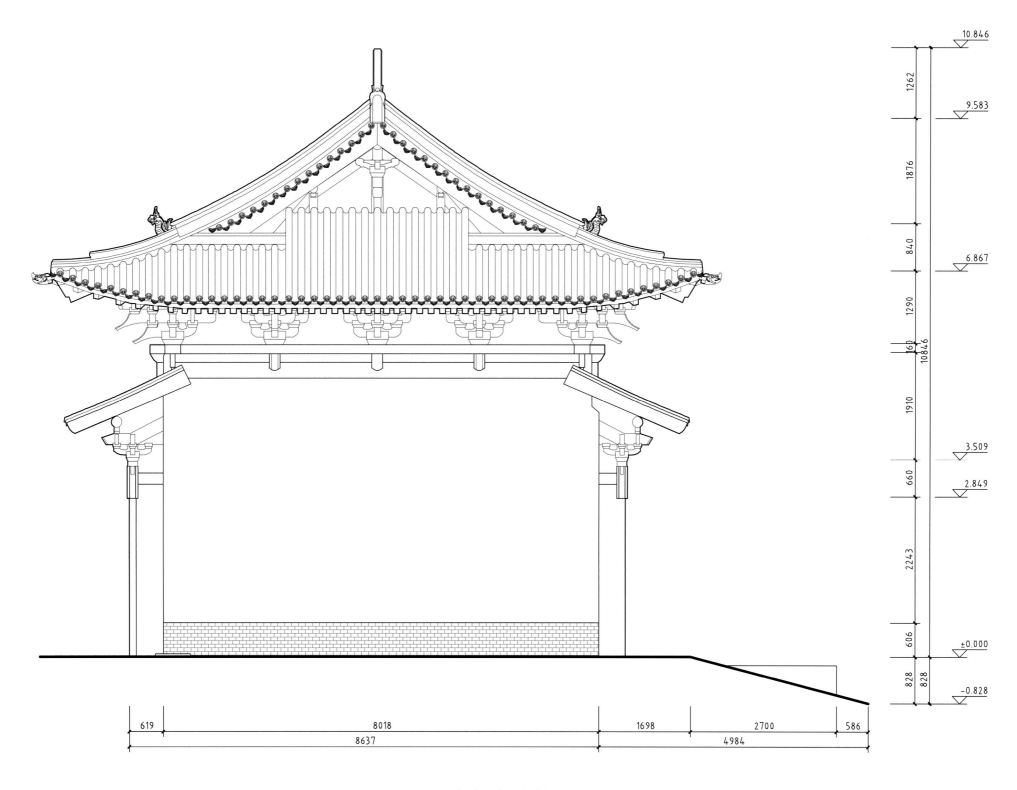

广胜下寺山门侧立面图
Side Elevation of the Hall of Heavenly Kings, Lower Guangsheng Monastery

0   0.5   1   2m

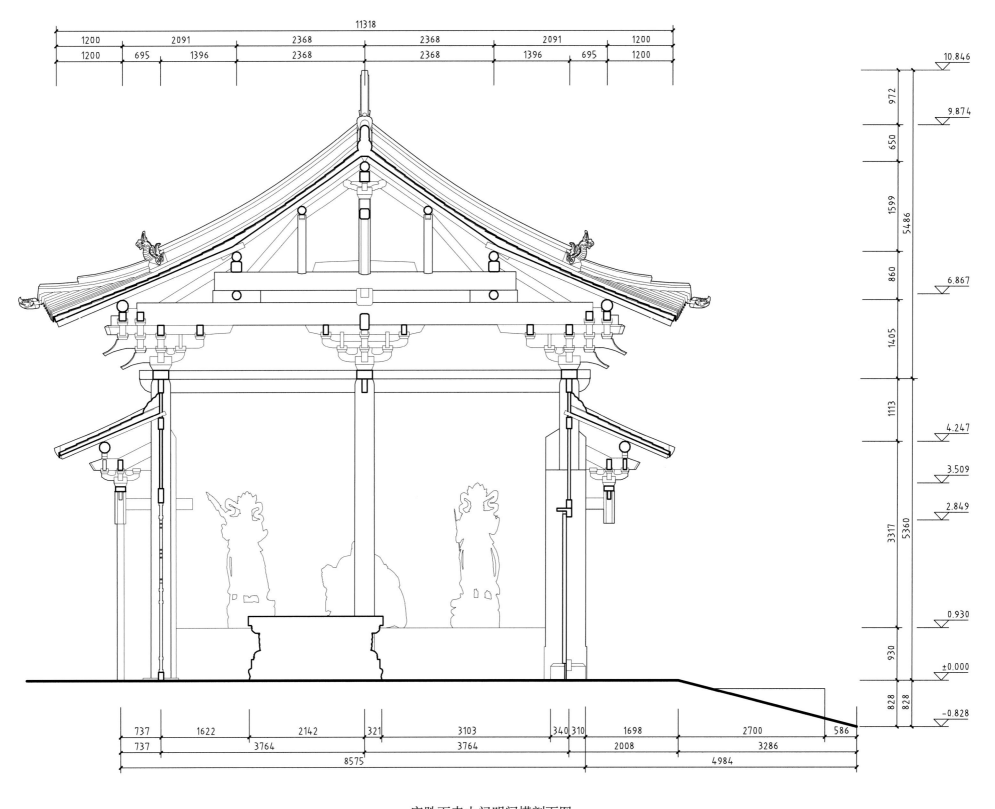

广胜下寺山门明间横剖面图
Cross Section of the Central Bay of the Hall of Heavenly Kings, Lower Guangsheng Monastery

0   0.5   1   2m

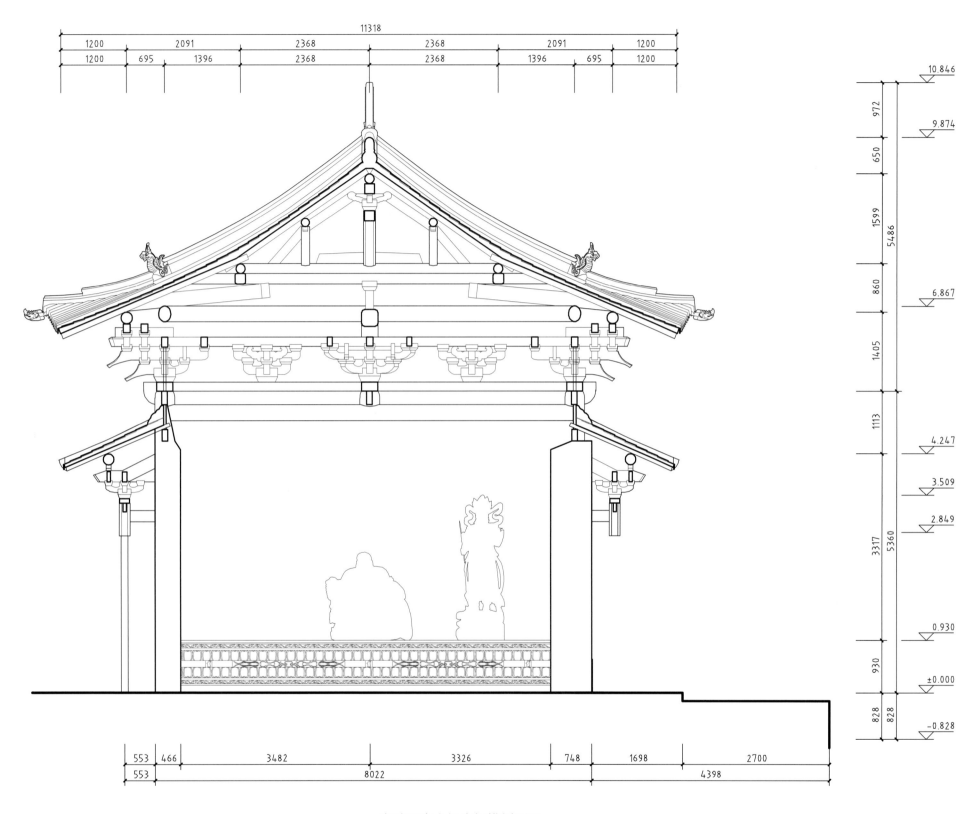

广胜下寺山门次间横剖面图

Cross Section of the Western and Eastern Bays of the Hall of Heavenly Kings, Lower Guangsheng Monastery

0 0.5 1 2m

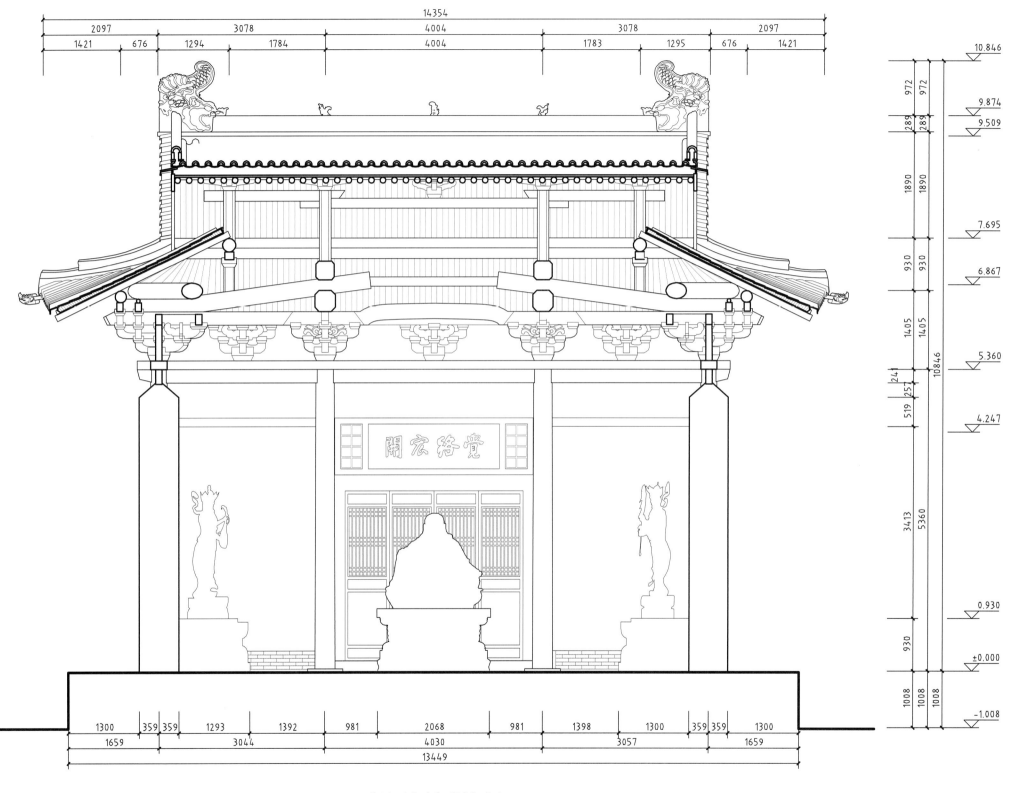

14354
2097　3078　4004　3078　2097
1421　676　1294　1784　4004　1783　1295　676　1421

10.846
972　972
289　289
9.874
9.509
1890　1890
7.695
930　930
6.867
1405　1405
5.360
241
257
519　257　10846
4.247
3413　5360

0.930
930
±0.000
1008　1008　1008
-1.008

觉路宏開

1300　359　359　1293　1392　981　2068　981　1398　1300　359　359　1300
1659　3044　4030　3057　1659
13449

广胜下寺山门纵剖面图
Longitudinal Section of the Hall of Heavenly Kings, Lower Guangsheng Monastery

0　0.5　1　2m

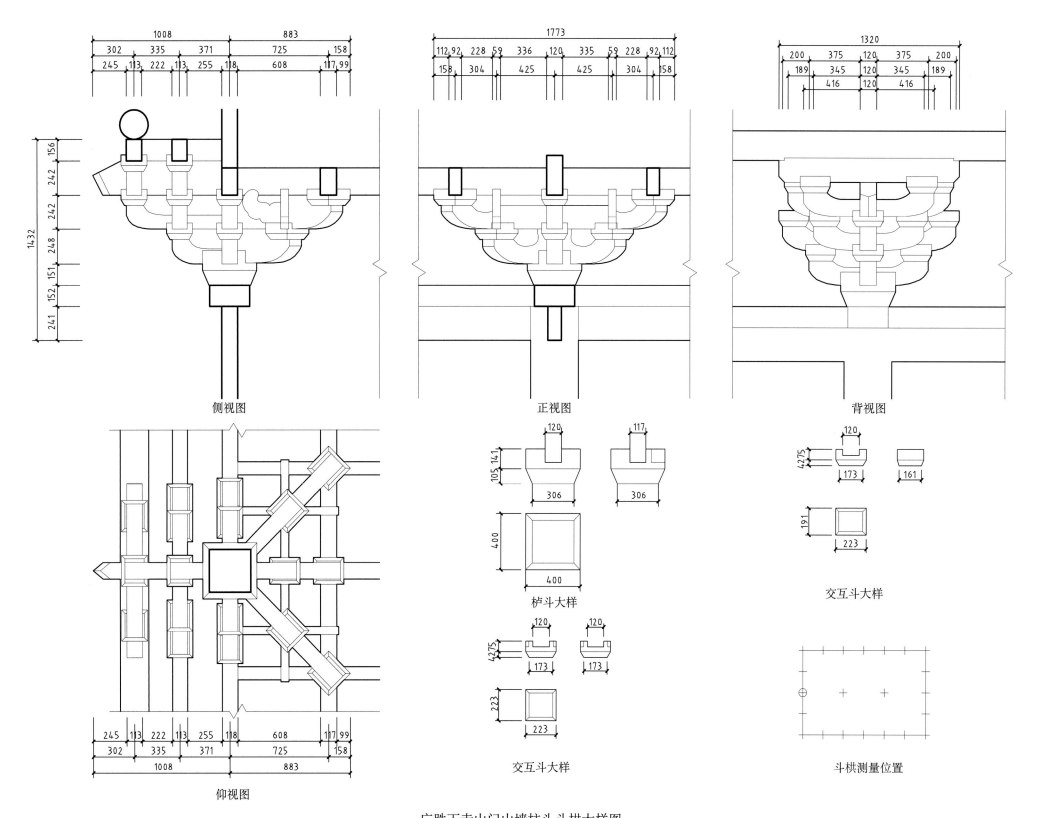

侧视图

正视图

背视图

仰视图

栌斗大样

交互斗大样

交互斗大样

斗栱测量位置

广胜下寺山门山墙柱头斗栱大样图

Detailed Drawing of Bracket Sets on Gable Walls' Columns of the Hall of Heavenly Kings, Lower Guangsheng Monastery

0   0.2   1m

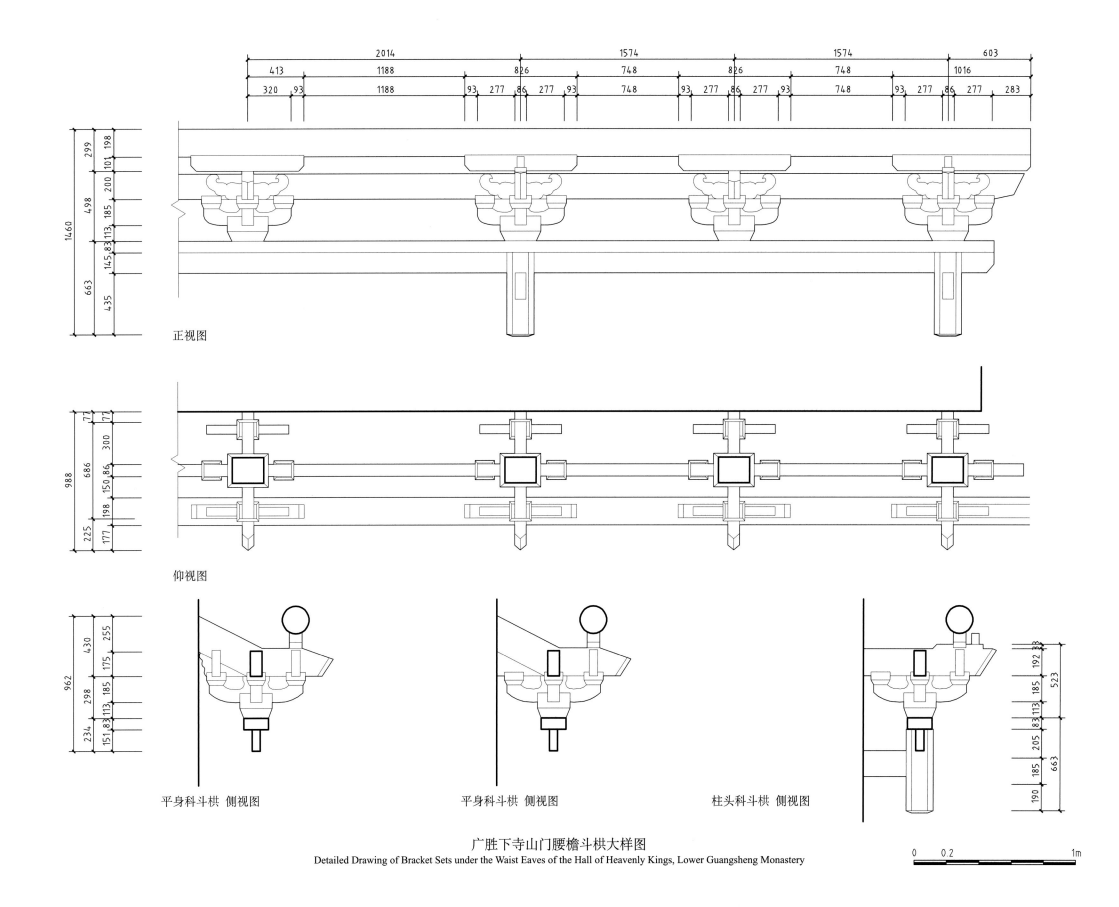

正视图

仰视图

平身科斗栱 侧视图

平身科斗栱 侧视图

柱头科斗栱 侧视图

广胜下寺山门腰檐斗栱大样图

Detailed Drawing of Bracket Sets under the Waist Eaves of the Hall of Heavenly Kings, Lower Guangsheng Monastery

0   0.2   1m

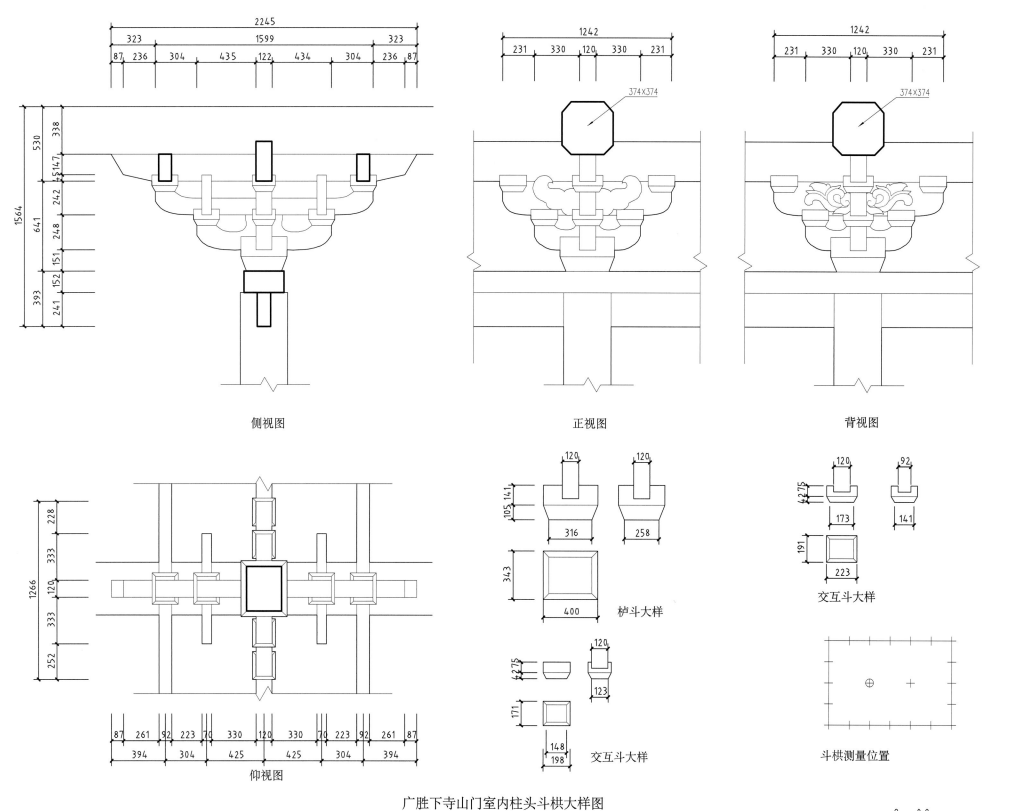

侧视图

正视图

背视图

仰视图

栌斗大样

交互斗大样

交互斗大样

斗栱测量位置

广胜下寺山门室内柱头斗栱大样图

Detailed Drawing of Bracket Sets on Interior Columns of the Hall of Heavenly Kings, Lower Guangsheng Monastery

0　0.2　　　　　1m

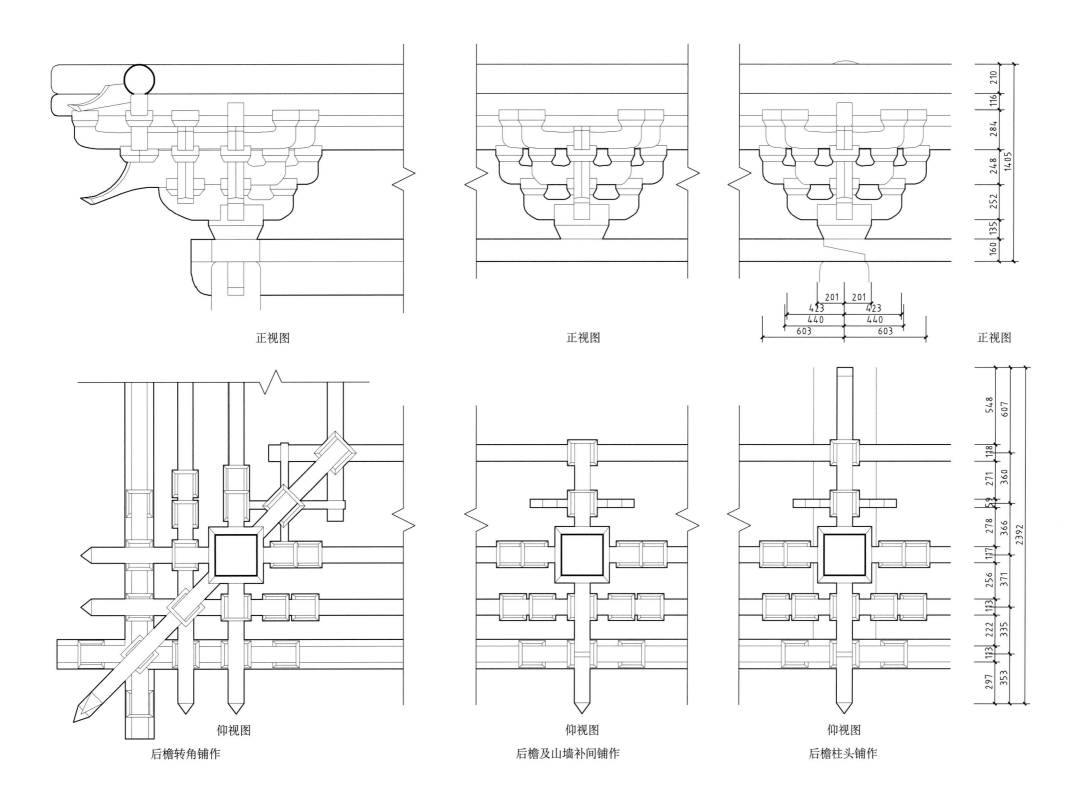

正视图

正视图

正视图

仰视图

仰视图

仰视图

后檐转角铺作

后檐及山墙补间铺作

后檐柱头铺作

广胜下寺山门后檐斗栱大样图（一）

Detailed Drawing of Bracket Sets on the Back Elevation of the Hall of Heavenly Kings, Lower Guangsheng Monastery (1)

0   0.2   1m

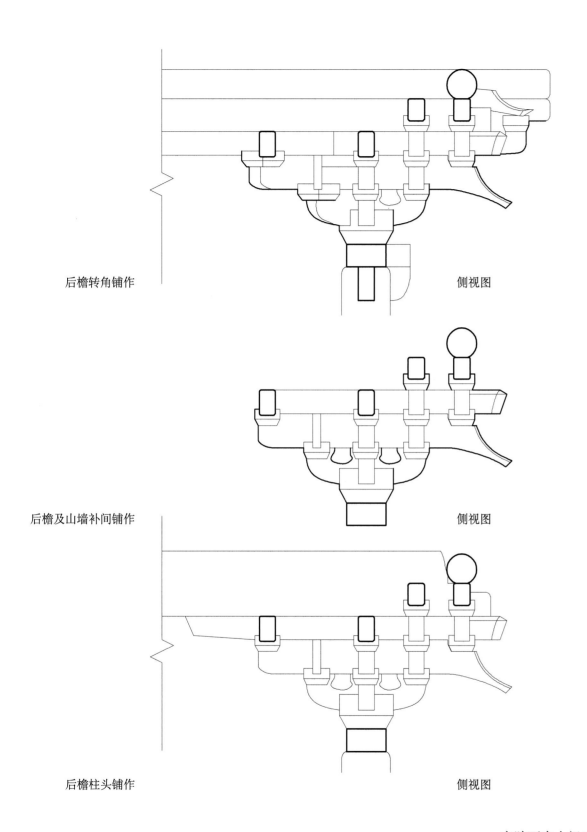

后檐转角铺作　　　　　　　　　　　　　　　侧视图

后檐及山墙补间铺作　　　　　　　　　　　　侧视图

后檐柱头铺作　　　　　　　　　　　　　　　侧视图

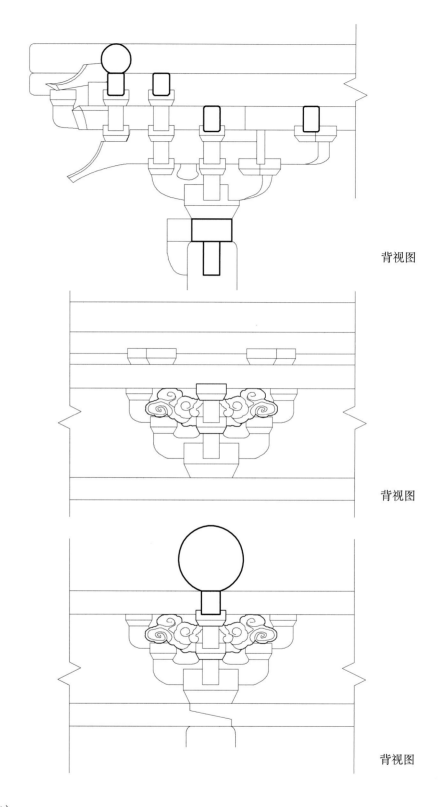

背视图

背视图

背视图

广胜下寺山门后檐斗栱大样图（二）

Detailed Drawing of Bracket Sets on the Back Elevation of the Hall of Heavenly Kings, Lower Guangsheng Monastery (2)

0　　0.2　　　　　　　1m

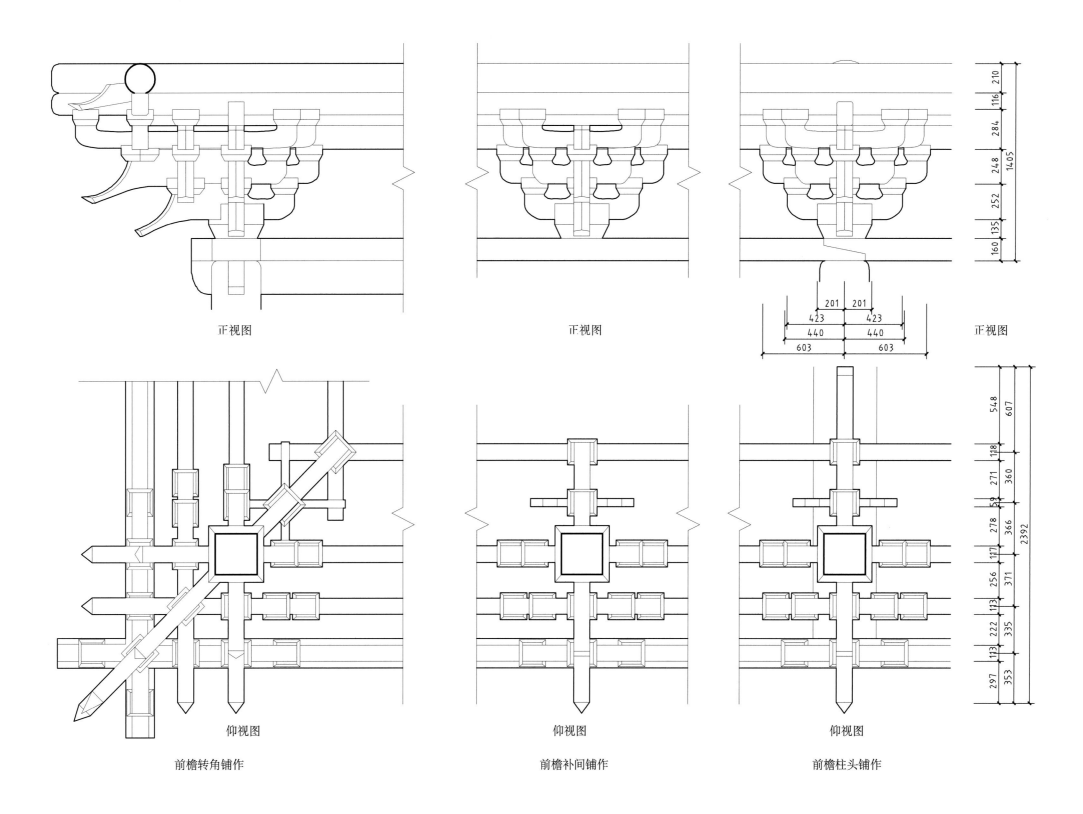

正视图　　　　　　　　　　正视图　　　　　　　　　　正视图

仰视图　　　　　　　　　　仰视图　　　　　　　　　　仰视图

前檐转角铺作　　　　　　　前檐补间铺作　　　　　　　前檐柱头铺作

广胜下寺山门前檐斗栱大样图（一）

Detailed Drawing of Bracket Sets on the Front Elevation of the Hall of Heavenly Kings, Lower Guangsheng Monastery (1)

0　0.2　　　　　　1m

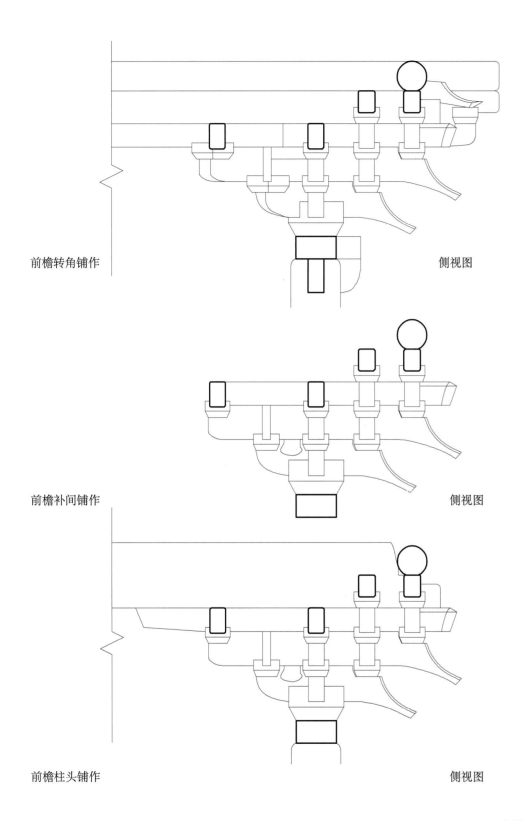

前檐转角铺作　　　　　　　　　　　　　　　　　　侧视图

前檐补间铺作　　　　　　　　　　　　　　　　　　侧视图

前檐柱头铺作　　　　　　　　　　　　　　　　　　侧视图

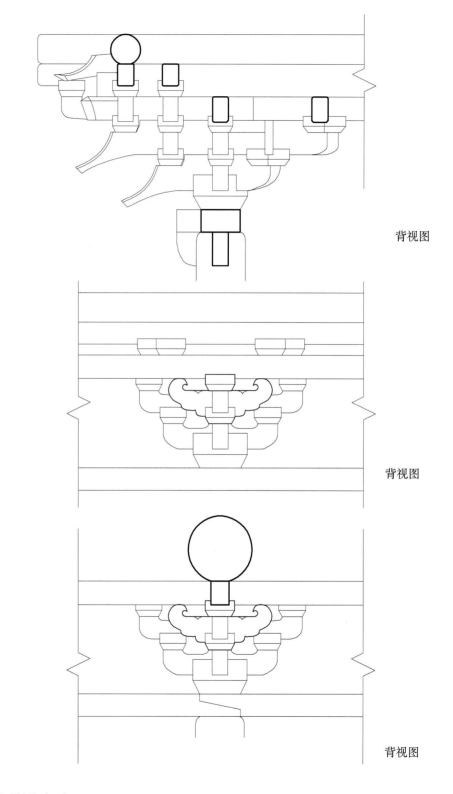

背视图

背视图

背视图

广胜下寺山门前檐斗栱大样图（二）

Detailed Drawing of Bracket Sets on the Front Elevation of the Hall of Heavenly Kings, Lower Guangsheng Monastery (2)

0　0.2　　　　1m

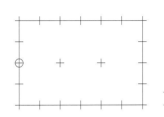

所测斗栱
位置标示（非腰檐斗栱）

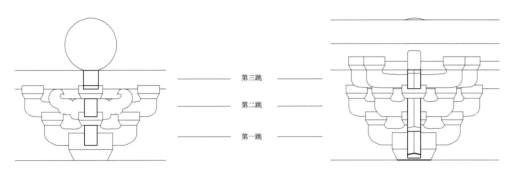

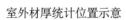

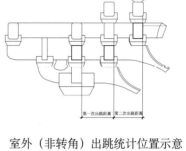

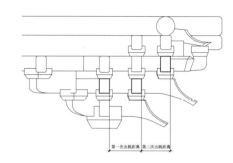

室内材厚统计位置示意　　　　室外材厚统计位置示意　　　　室外（非转角）出跳统计位置示意　　　　转角铺作出跳统计位置示意

183

室外材厚测量统计

斗栱		第一跳（前檐为假昂，后檐为华栱）	第二跳（昂）	第三跳（耍头）
1号	面阔方向	121	116	
	斜向	115	117	
	进深方向	114	120	117
2号		119	115	120
3号		114	118	122
4号		117	115	122
5号		120	118	118
6号		119	117	119
7号	面阔方向	115	112	
	斜向	117	118	119
	进深方向	115	119	119
8号	面阔方向	115	115	
	斜向	116	117	117
	进深方向	113	111	119
9号		118	120	122
10号		117	110	125
11号		120	117	104
12号		113	112	120
13号		112	115	112
14号	面阔方向	120	112	
	斜向	120	115	
	进深方向	123	200	120

室内材厚测量统计

斗栱	第一跳（华栱）	第二跳（华栱）	第三跳（蚂蚱头）
2号	116	120	
3号	116	117	134
4号	114	120	
5号	117	118	130
6号	112	123	
9号	119	116	
10号	122	115	141
11号	121	124	
12号	118	120	144
13号	116	120	

室外（非转角）出跳统计

斗栱	第一次出跳距离	第二次出跳距离
2号	376	333
3号	364	342
4号	382	347
5号	369	348
6号	353	332
8号	376	354
10号	378	330
11号	356	343
12号	353	333
13号	381	334
20号	394	326
平均值	371	338

转角铺作部分栱间距统计

斗栱	第一次出跳距离	第二次出跳距离
7号垂直山墙栱距	343	324
7号垂直正立面栱距	385	337
8号垂直正立面栱距	363	323
14号垂直正立面栱距	367	360
14号垂直山墙栱距	377	312
平均值	367	331

广胜下寺山门斗栱数据统计表（一）
Statistics of Bracket Sets of the Hall of Heavenly Kings, Lower Guangsheng Monastery (1)

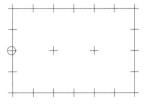

所测斗栱
位置标示（非腰檐斗栱）

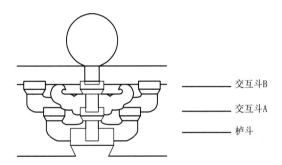

交互斗B
交互斗A
栌斗

轴线上斗尺寸统计位置示意

数理统计分析汇总			
材厚	蚂蚱头	平均值	137.25
		标准差	6.40
	华栱（昂、假昂）	平均值	117.25
		标准差	3.20
		材厚95%置信区间	116.52～117.98
栌斗	栌斗下底	平均值	306.50
		标准差	6.36
	栌斗上边	平均值	401.90
		标准差	3.11
交互斗	交互斗底边	平均值	169.80
		标准差	3.12
	交互斗上边	平均值	220.85
		标准差	7.62
出跳	第一次出跳	平均值	368.71
		标准差	13.78
	第二次出跳	平均值	334.80
		标准差	12.42

注：
1. 本次统计中，材厚数据、斗尺寸数据来自实地手工测量，出跳距离来自三维扫描点云图像的测量。
2. 由于转角铺作手工测量有一定难度，三维点云图像因遮挡而有遗漏，所以在转角铺作的测量数据有所欠缺。
3. 在计算材厚平均数时，斗栱14进深方向第二跳（下昂）数据与其他数据偏离过大，舍弃之。斗栱11室外测量耍头数据亦有较大偏离，怀疑为严重毁坏，亦弃之。
4. 材厚的数据按照其服从正态分布的假定处理。利用矩法估计其参数值，有 $\mu=117.25$，$\sigma=3.199$，该估计为无偏估计。
5. 材厚的概率分布大致可用下方公式拟合：$f(x)=1/8.02 \times \exp\{-(x-117.25)^2/20.47\}$。
6. 材厚置信区间含义为，真实材厚值落在116.52～117.98范围内的概率为95%。
7. 斗栱材厚117左右，约合3.51寸，足材广在261左右。故对照宋《营造法式》对七等材的记载：高5.25寸，厚3.5寸，用于小殿或亭榭。测量尺寸与《营造法式》尺寸如此接近，说明建成年代应当在宋及宋之后木构体系相当成熟的阶段。
8. 斗栱材厚117，约合3.51寸，在清代斗口制中为六等材。
9. 本测量结果数据未注明单位者，均以mm作为单位。

轴线位置斗室内测量统计						
斗栱	栌斗底边	栌斗上边	交互斗A底边	交互斗A上边	交互斗B底边	交互斗B上边
2号	308	401	165	218	169	223
3号	310	407	168	225	176	222
4号	315	406	171	244	168	220
5号	312	399	172	222	168	217
6号	314	403	173	216	167	223
9号	300	401	169	218	165	219
10号	305	404	170	215	168	206
11号	298	401	174	221	166	222
12号	298	400	174	222	169	212
13号	305	397	171	221	173	221

广胜下寺山门斗栱数据统计表（二）
Statistics of Bracket Sets of the Hall of Heavenly Kings, Lower Guangsheng Monastery (2)

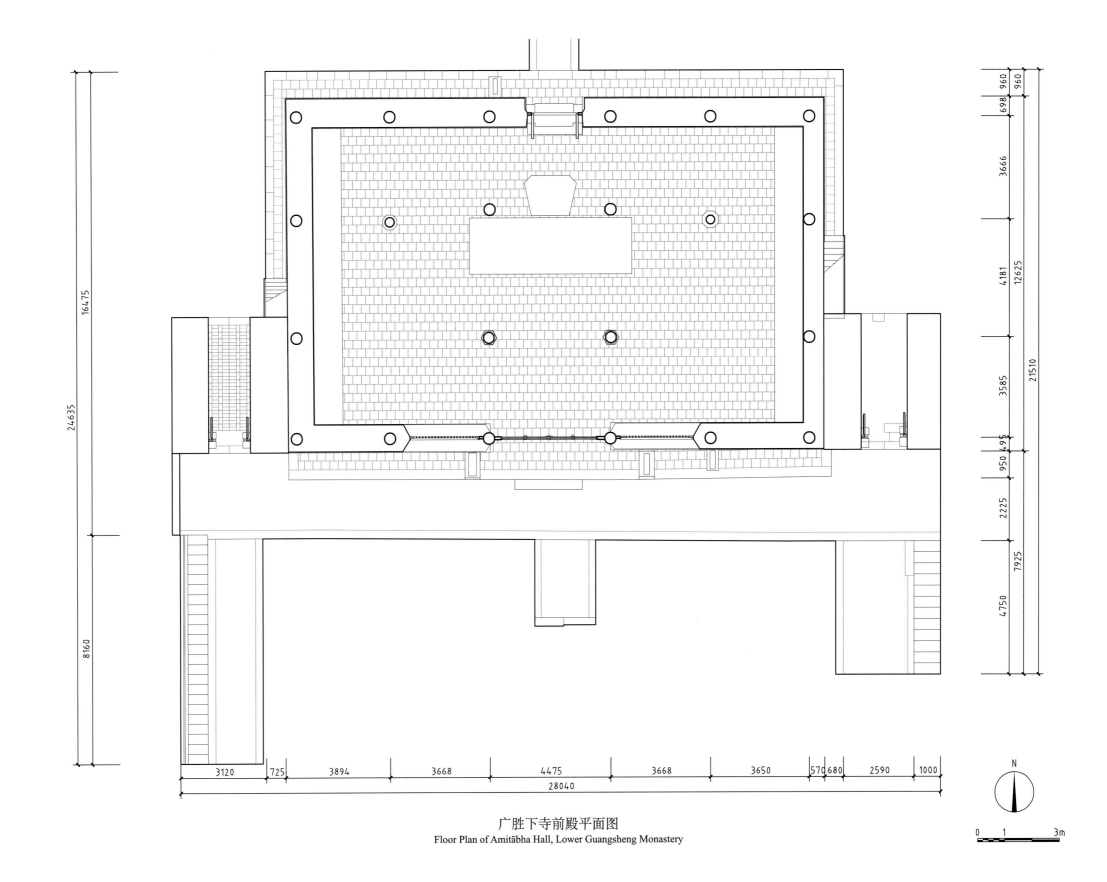

广胜下寺前殿平面图
Floor Plan of Amitābha Hall, Lower Guangsheng Monastery

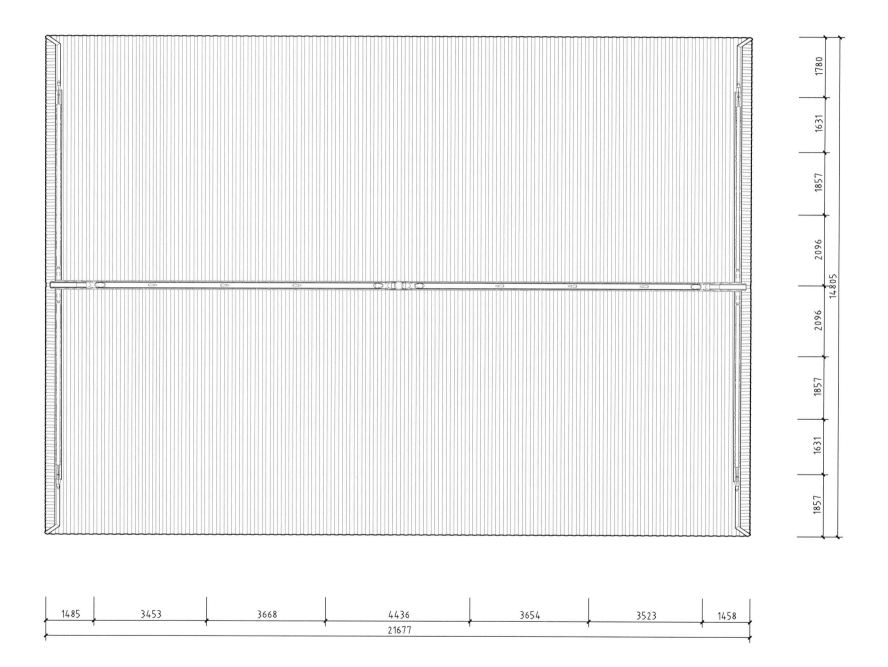

广胜下寺前殿屋顶平面图
Top View of the Roof of Amitābha Hall, Lower Guangsheng Monastery

0 1 3m

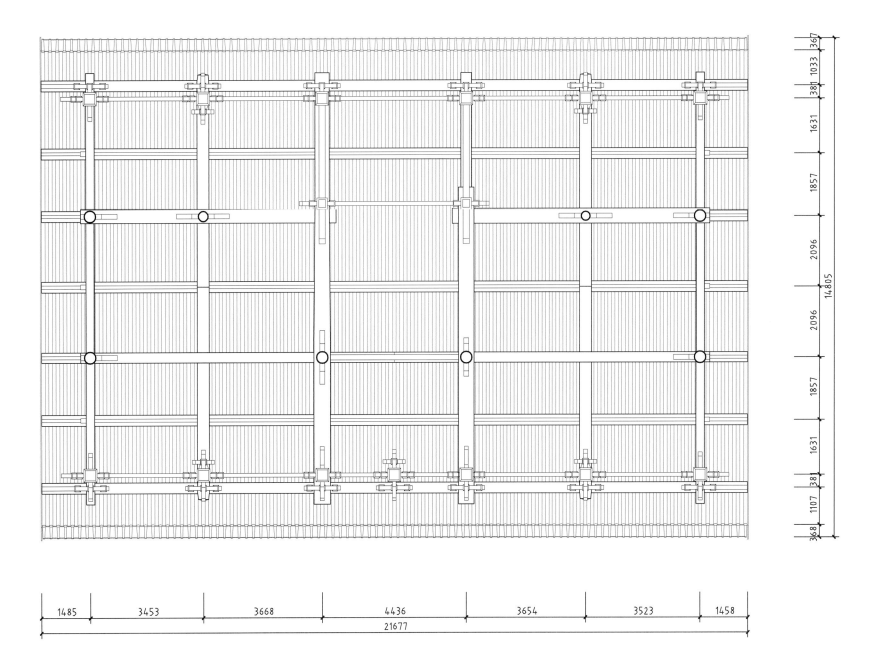

广胜下寺前殿梁架仰视平面图
Bottom View of the Truss of Amitābha Hall, Lower Guangsheng Monastery

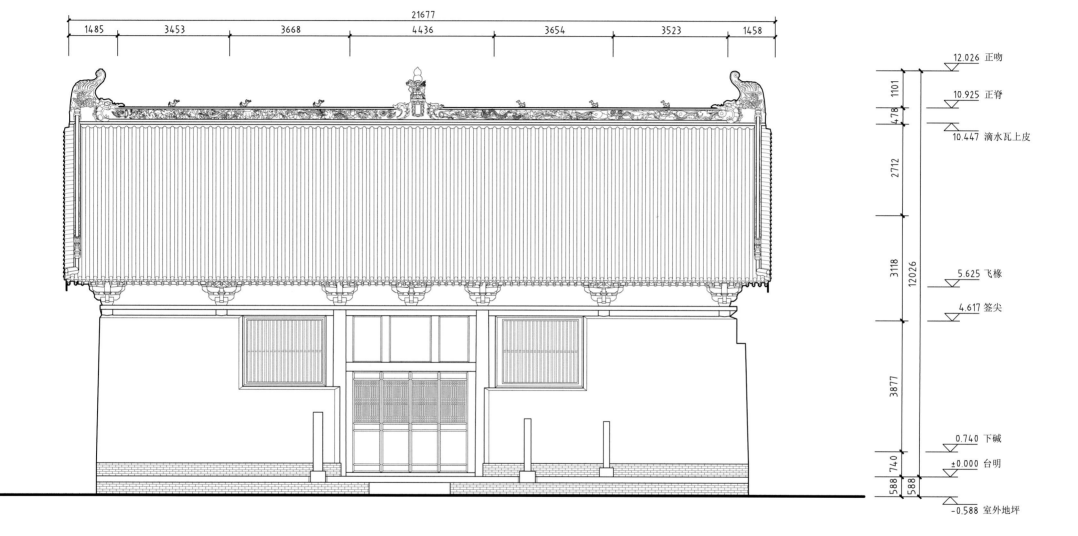

广胜下寺前殿南立面图
Southern Elevation of Amitābha Hall, Lower Guangsheng Monastery

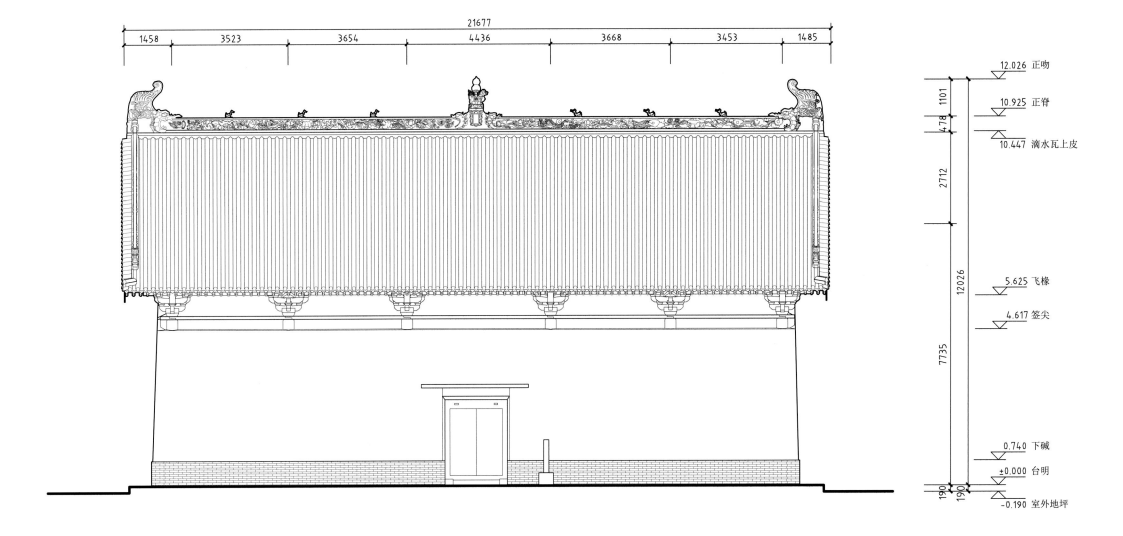

21677

| 1458 | 3523 | 3654 | 4436 | 3668 | 3453 | 1485 |

12.026 正吻

1101

10.925 正脊

4.78

10.447 滴水瓦上皮

2712

12026

5.625 飞椽

4.617 箅尖

7735

0.740 下碱

±0.000 台明

190 190

-0.190 室外地坪

| 461 | 3650 | 3668 | 4475 | 3668 | 3586 | 412 |

19920

广胜下寺前殿北立面图
Northern Elevation of Amitābha Hall, Lower Guangsheng Monastery

0 1 3m

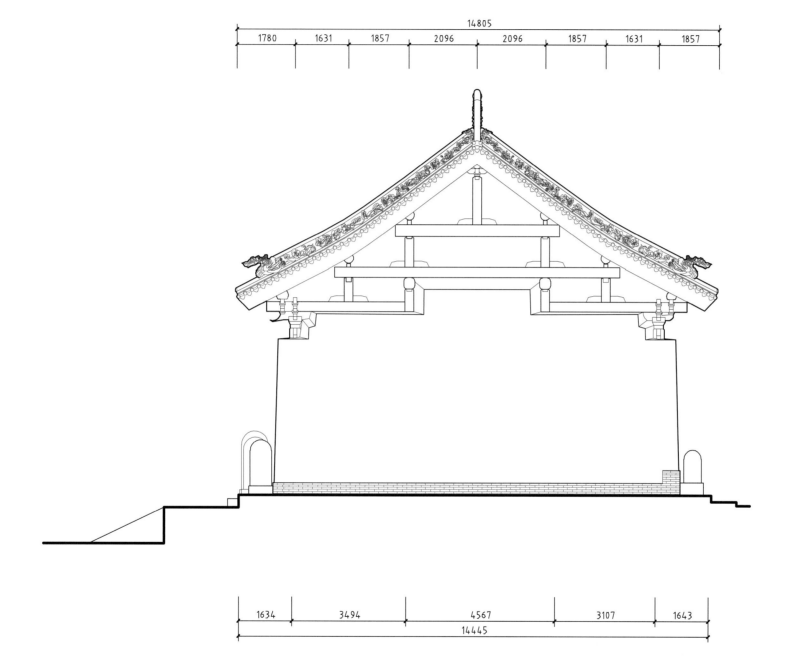

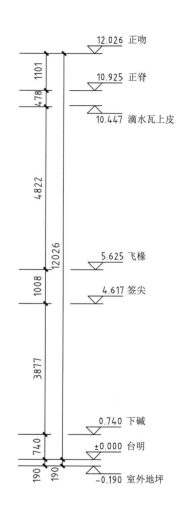

14805

1780 | 1631 | 1857 | 2096 | 2096 | 1857 | 1631 | 1857

1634 | 3494 | 4567 | 3107 | 1643

14445

12.026 正吻
10.925 正脊
10.447 滴水瓦上皮
5.625 飞椽
4.617 签尖
0.740 下碱
±0.000 台明
-0.190 室外地坪

1101
4.78
4822
12026
1008
3877
740
190 | 190

广胜下寺前殿东立面图
Eastern Elevation of Amitābha Hall, Lower Guangsheng Monastery

0   1   3m

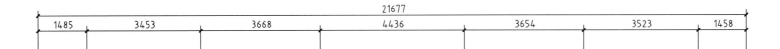

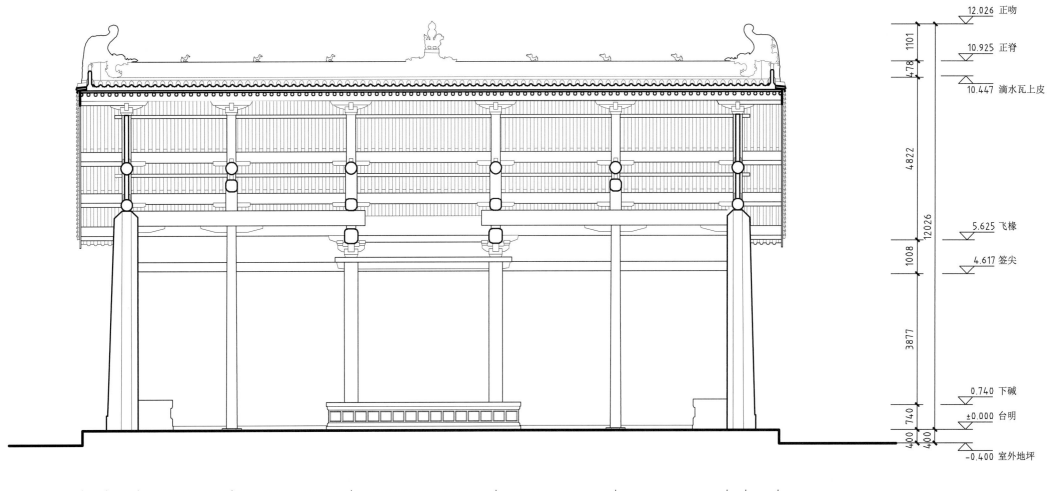

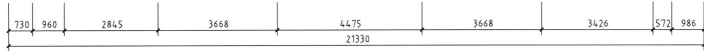

广胜下寺前殿纵剖面图
Longitudinal Section of Amitābha Hall, Lower Guangsheng Monastery

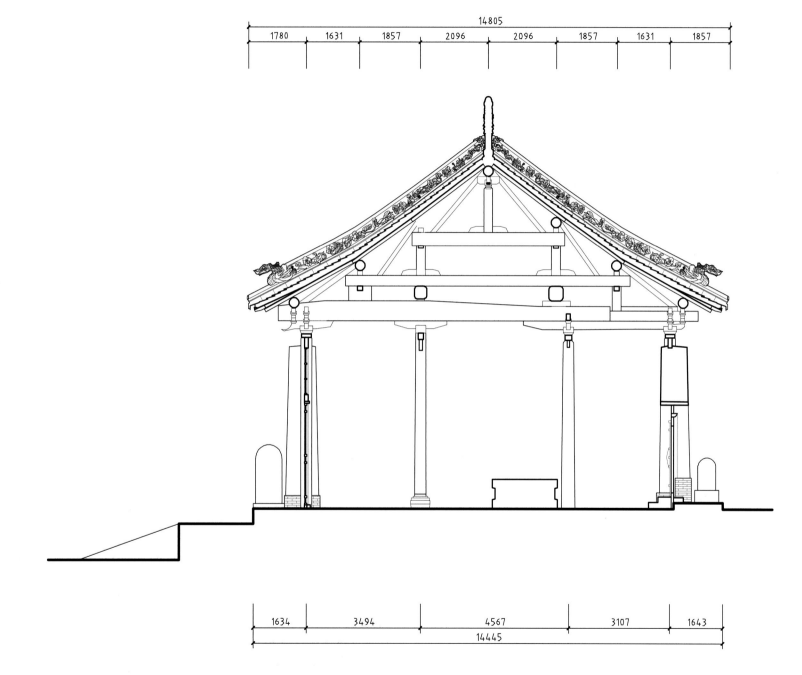

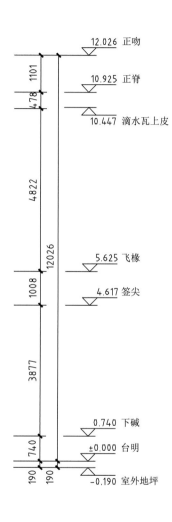

14805

| 1780 | 1631 | 1857 | 2096 | 2096 | 1857 | 1631 | 1857 |

12.026 正吻
1101
10.925 正脊
478
10.447 滴水瓦上皮

4822

12026

5.625 飞椽
1008
4.617 签尖

3877

0.740 下碱
740
±0.000 台明
190 190
-0.190 室外地坪

| 1634 | 3494 | 4567 | 3107 | 1643 |

14445

广胜下寺前殿明间横剖面图
Cross Section of the Central Bay of Amitābha Hall, Lower Guangsheng Monastery

0 1 3m

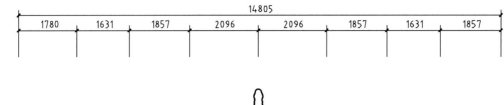

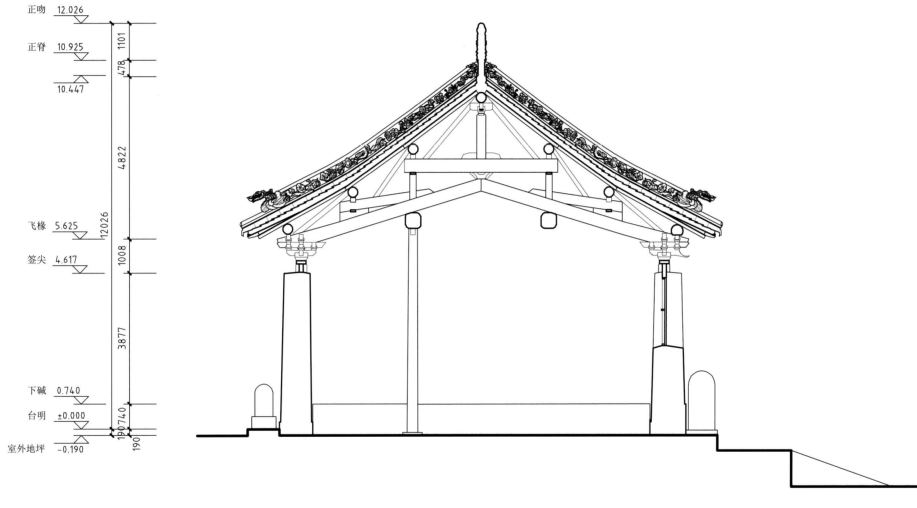

正吻 12.026

正脊 10.925

10.447

飞椽 5.625

签尖 4.617

下碱 0.740

台明 ±0.000

室外地坪 -0.190

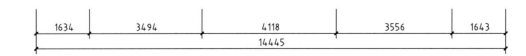

广胜下寺前殿次间横剖面图
Cross Section of the Bay next to the Central Bay of Amitābha Hall, Lower Guangsheng Monastery

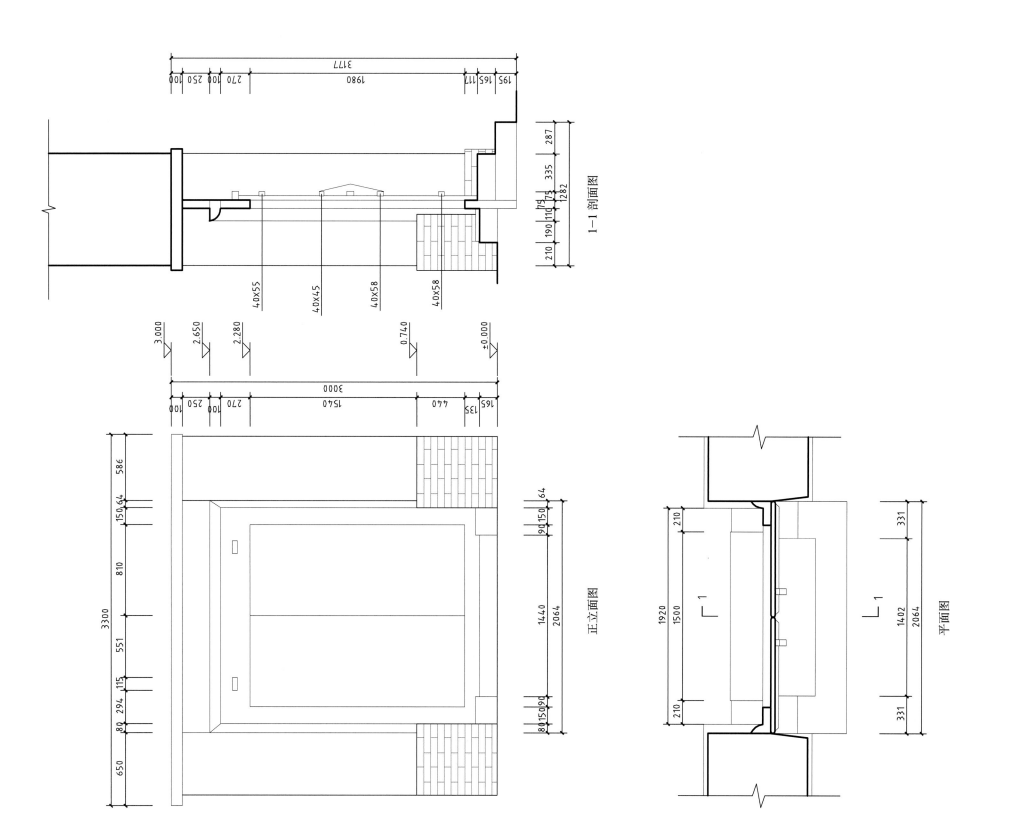

1—1 剖面图

正立面图

平面图

广胜下寺前殿北门大样图
Detailed Drawing of the North Gate of Amitābha Hall, Lower Guangsheng Monastery

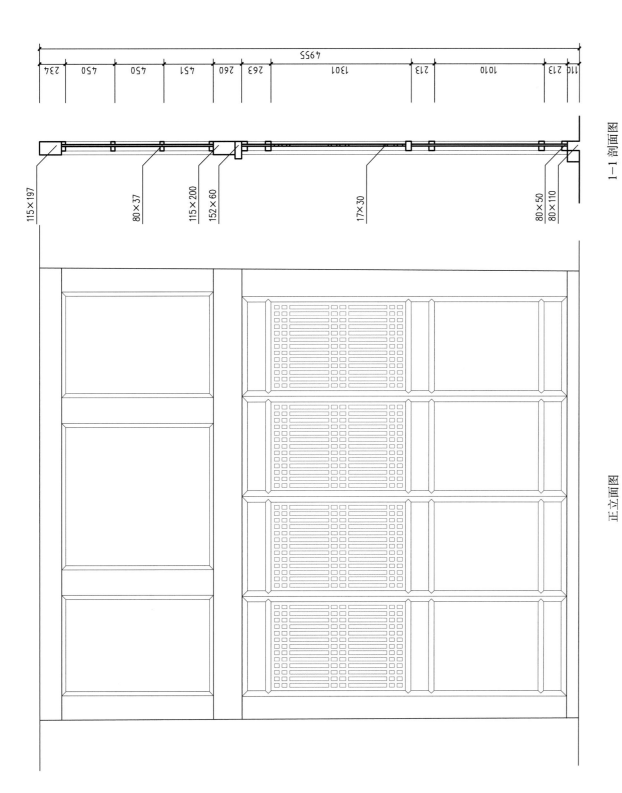

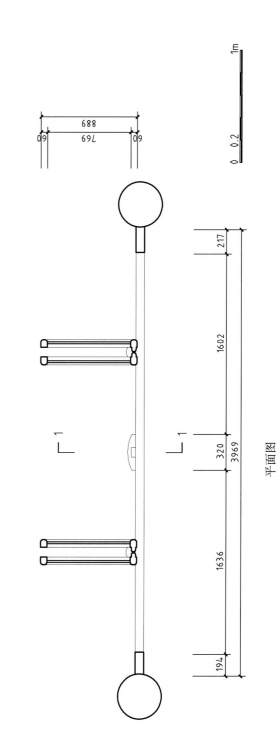

正立面图

1—1 剖面图

平面图

广胜下寺前殿南门大样图

Detailed Drawing of the South Gate of Amitābha Hall, Lower Guangsheng Monastery

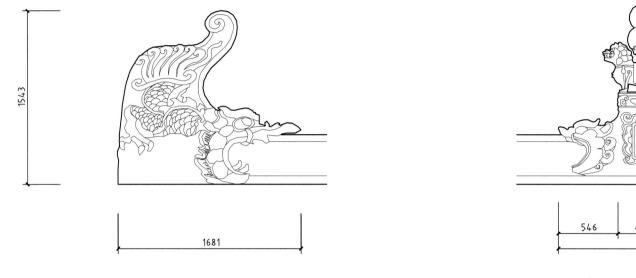

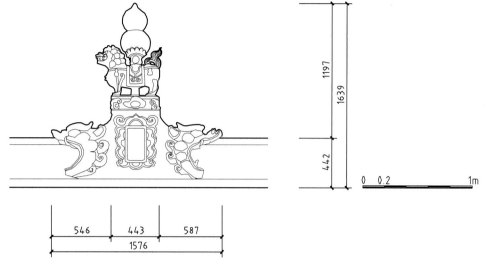

0 0.2   1m

广胜下寺前殿屋脊大样图
Detailed Drawing of the Main Ridge of Amitābha Hall, Lower Guangsheng Monastery

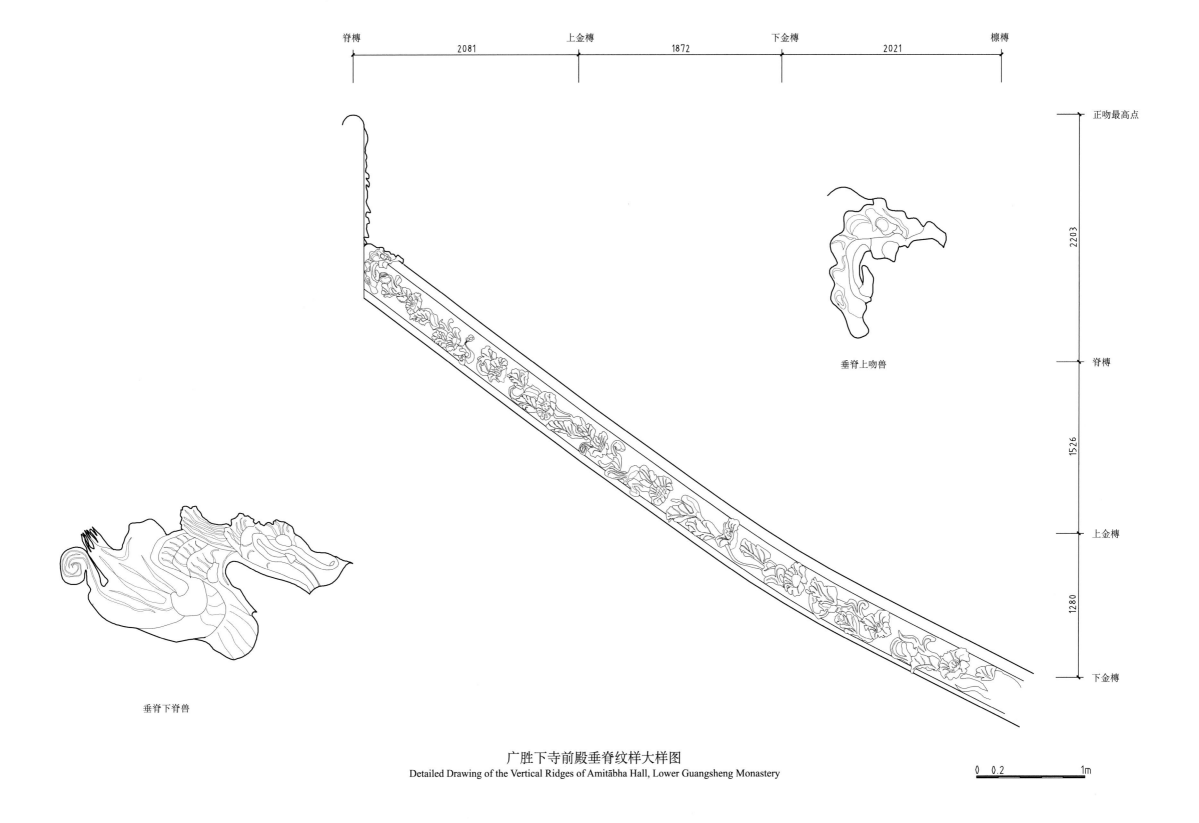

脊槫　　　2081　　　上金槫　　　1872　　　下金槫　　　2021　　　檐槫

正吻最高点

2203

脊槫

1526

上金槫

1280

下金槫

垂脊上吻兽

垂脊下脊兽

广胜下寺前殿垂脊纹样大样图

Detailed Drawing of the Vertical Ridges of Amitābha Hall, Lower Guangsheng Monastery

0　0.2　　　1m

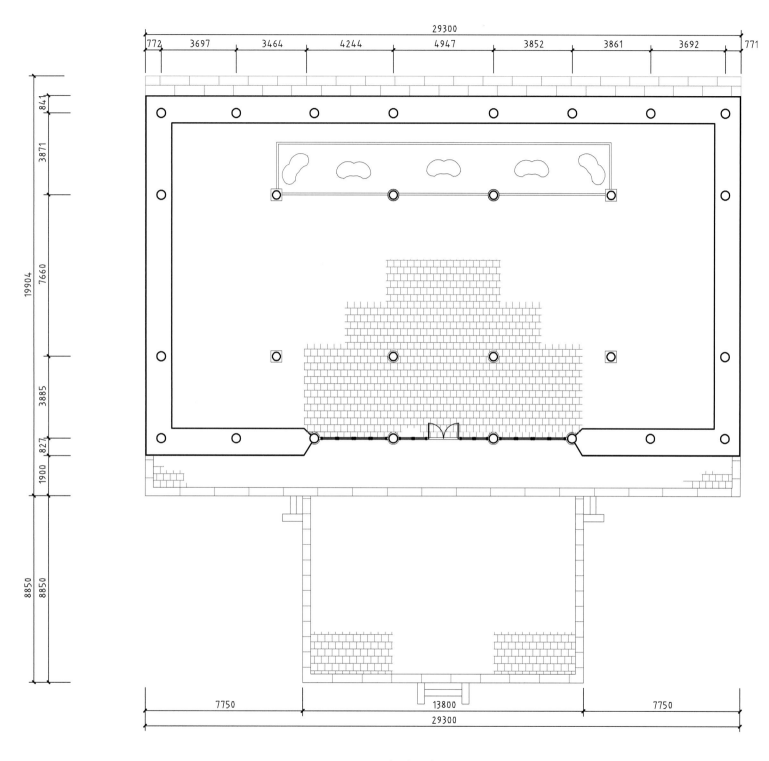

广胜下寺后殿平面图
Floor Plan of the Great Buddha Hall, Lower Guangsheng Monastery

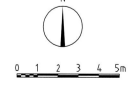

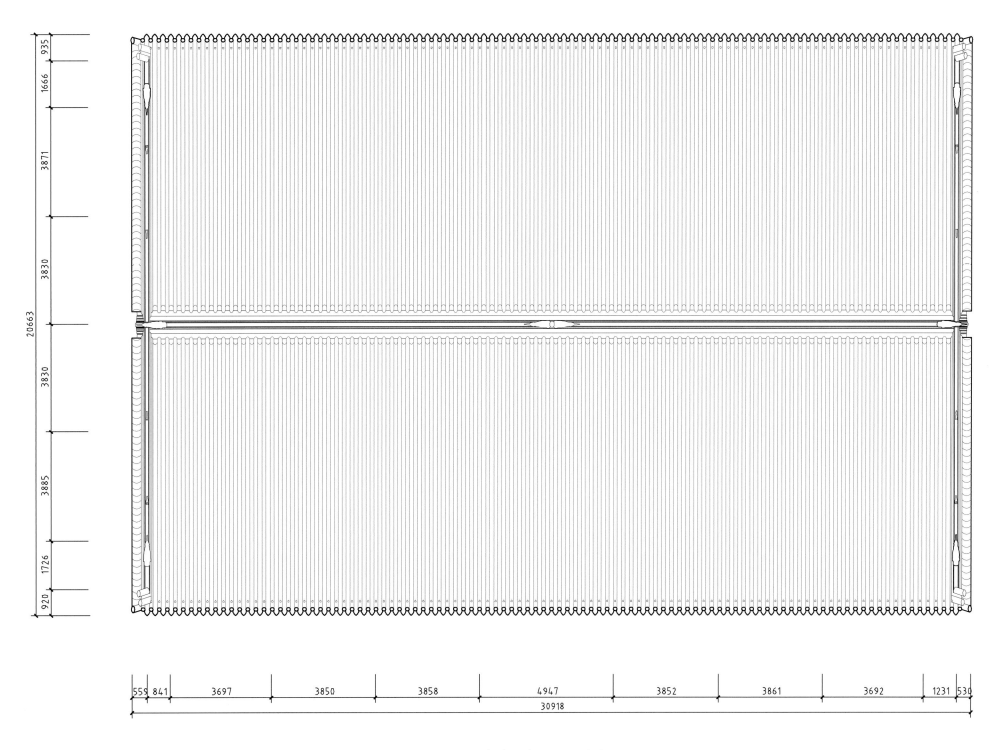

广胜下寺后殿屋顶平面图

Top View of the Roof of the Great Buddha Hall, Lower Guangsheng Monastery

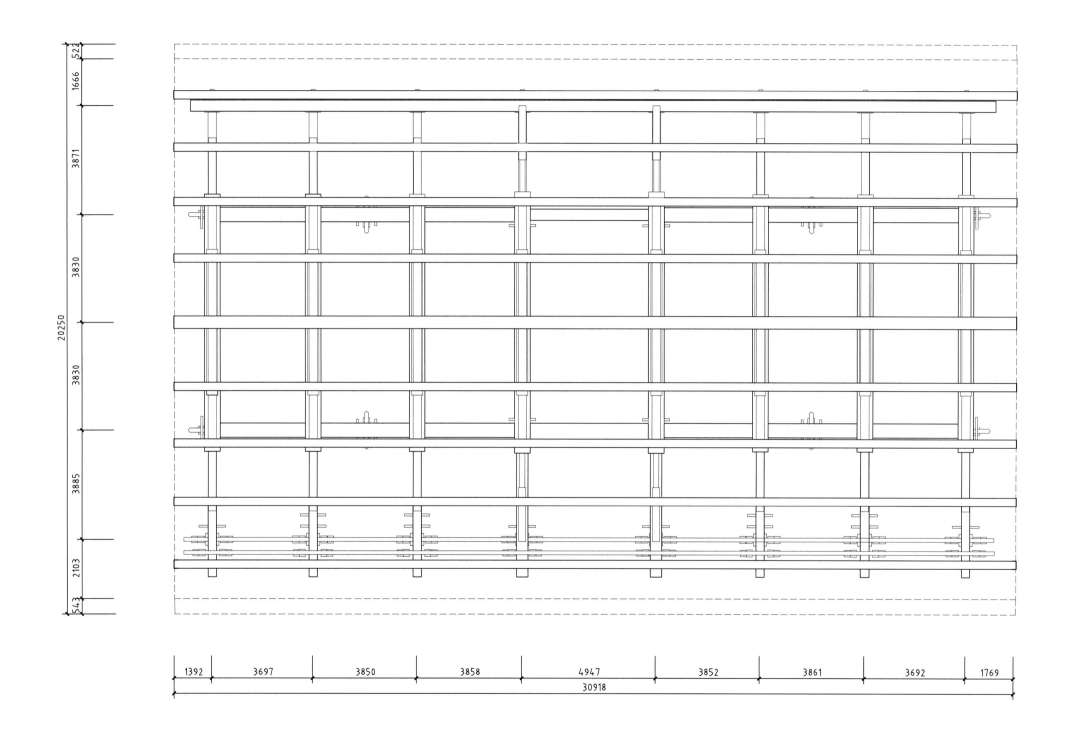

广胜下寺后殿梁架俯视平面图
Top View of the Truss of the Great Buddha Hall, Lower Guangsheng Monastery

0   1   3m

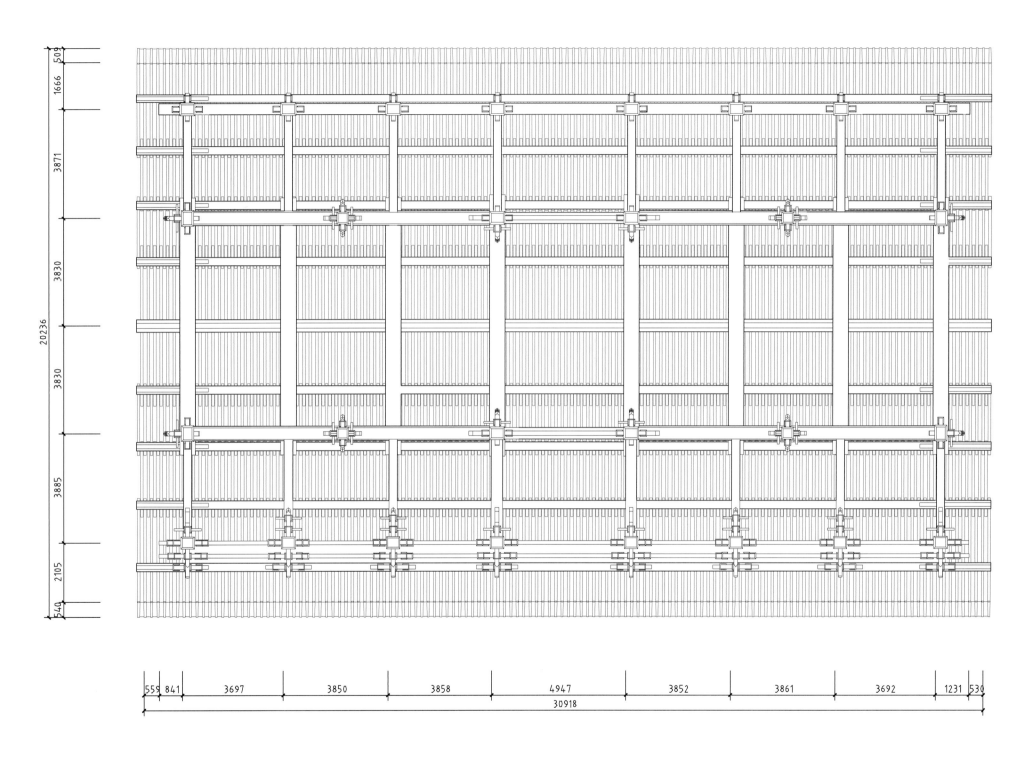

广胜下寺后殿梁架仰视平面图
Bottom View of the Truss of the Great Buddha Hall, Lower Guangsheng Monastery

0 1 3m

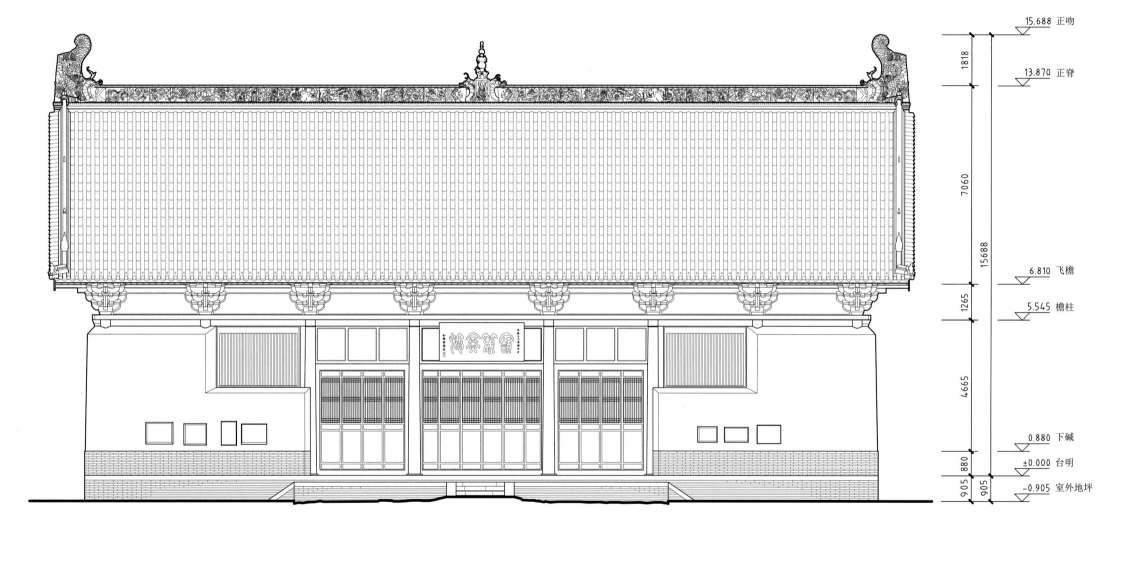

15.688 正吻

13.870 正脊

6.810 飞檐

5.545 檐柱

0.880 下碱

±0.000 台明

-0.905 室外地坪

1818

7060

15688

1265

4665

880

905

905

772 3697 3848 3860 4947 3852 3861 3692 771

29300

广胜下寺后殿南立面图

Southern Elevation of the Great Buddha Hall, Lower Guangsheng Monastery

0 1 3m

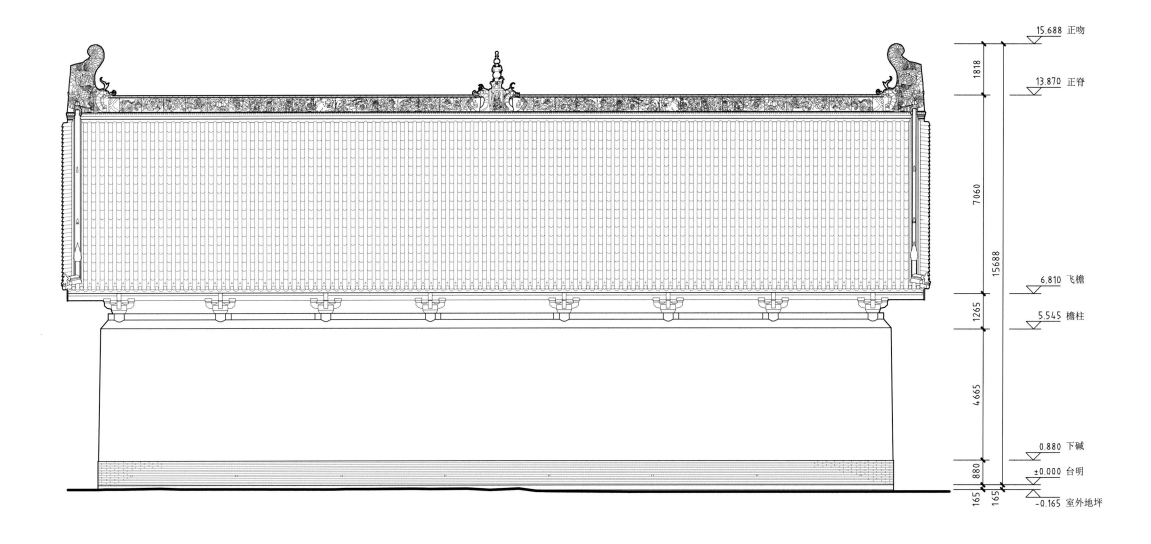

15.688 正吻

13.870 正脊

6.810 飞檐

5.545 檐柱

0.880 下碱

±0.000 台明

-0.165 室外地坪

1818

7060

15688

1265

4665

880

165

165

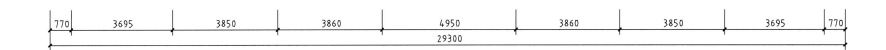

770　3695　3850　3860　4950　3860　3850　3695　770

29300

广胜下寺后殿北立面图
Northern Elevation of the Great Buddha Hall, Lower Guangsheng Monastery

0　1　3m

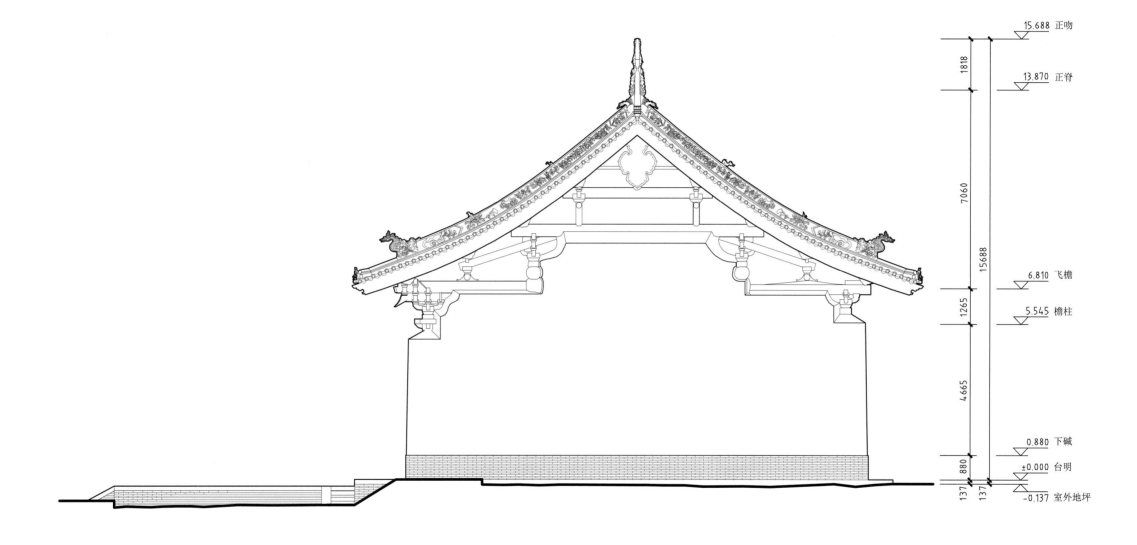

15.688 正吻

13.870 正脊

6.810 飞檐

5.545 檐柱

0.880 下碱

±0.000 台明

-0.137 室外地坪

1818

7060

15688

1265

4665

880

137

137

1000   7630   350   2770   841   3871   3830   3830   3885   827

11750   17084

广胜下寺后殿东立面图
Eastern Elevation of the Great Buddha Hall, Lower Guangsheng Monastery

0   1   3m

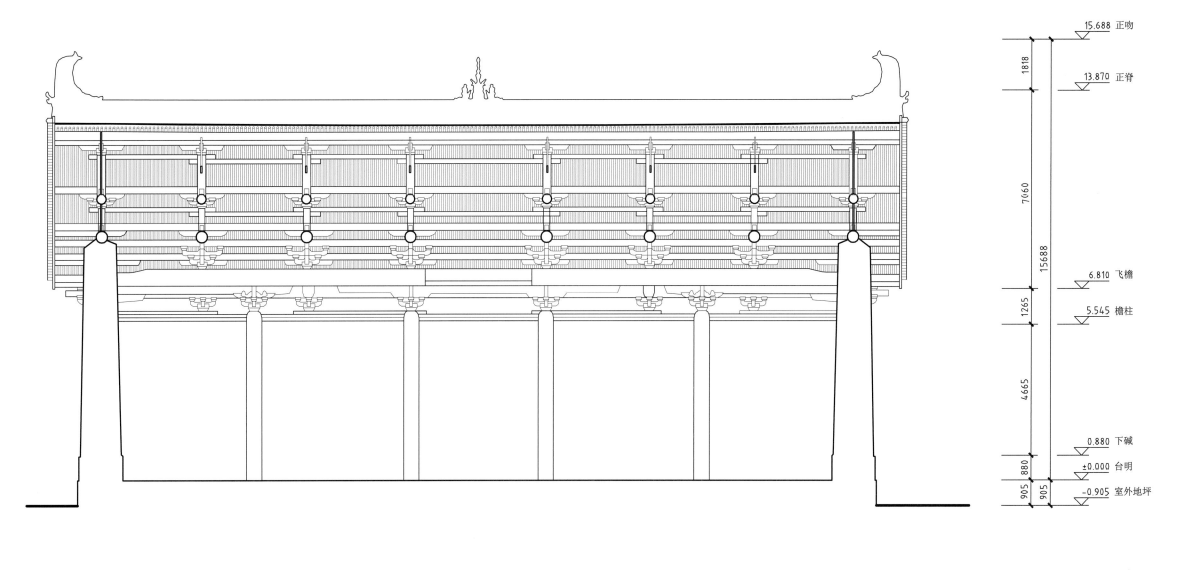

15.688 正吻
13.870 正脊
6.810 飞檐
5.545 檐柱
0.880 下碱
±0.000 台明
-0.905 室外地坪

1818
7660
15688
1265
4665
880
905
905

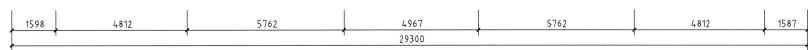

1598　4812　5762　4967　5762　4812　1587
29300

广胜下寺后殿纵剖面图
Longitudinal Section of the Great Buddha Hall, Lower Guangsheng Monastery

0　1　3m

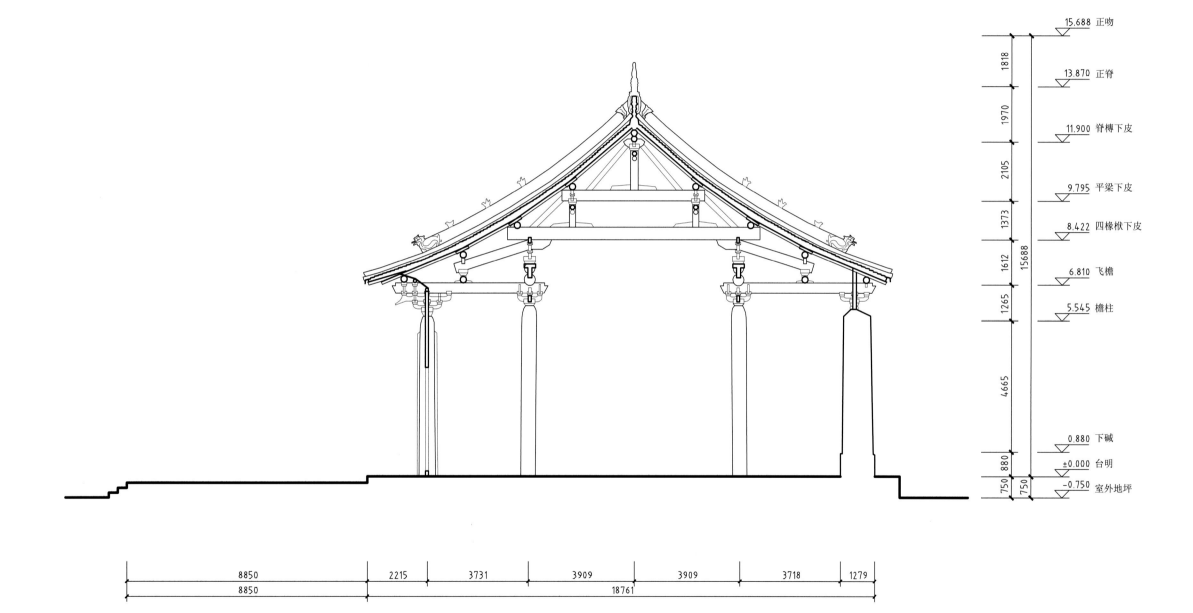

15.688 正吻

1818

13.870 正脊

1970

11.900 脊槫下皮

2105

9.795 平梁下皮

1373

8.422 四椽栿下皮

1612

6.810 飞檐

1265

5.545 檐柱

4665

15688

0.880 下碱

880

±0.000 台明

750

750

-0.750 室外地坪

8850　2215　3731　3909　3909　3718　1279
8850　18761

广胜下寺后殿明间横剖面图
Cross Section of the Central Bay of the Great Buddha Hall, Lower Guangsheng Monastery

0　1　3m

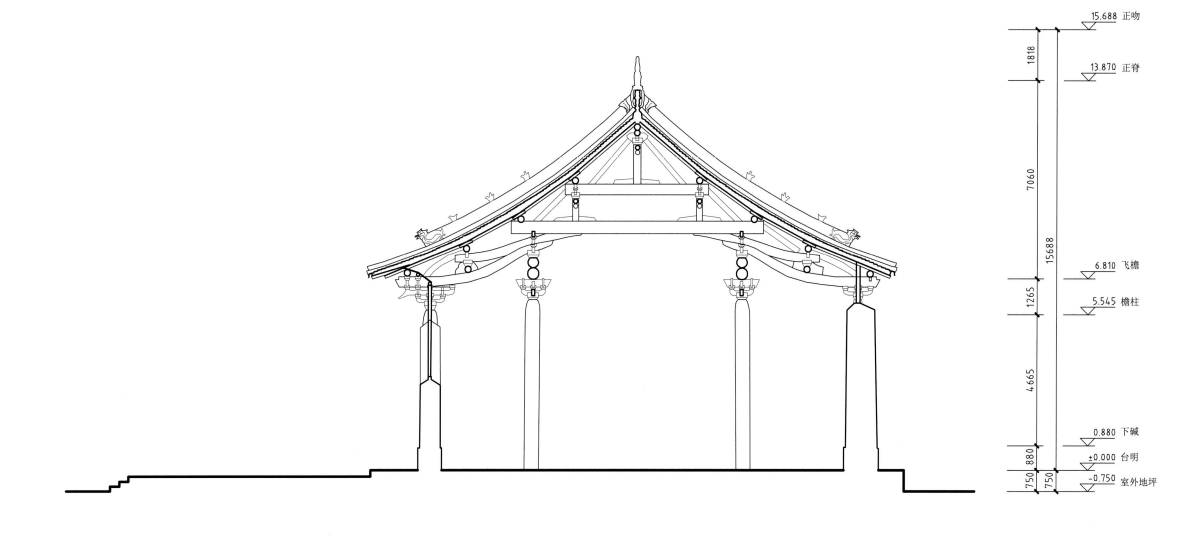

15.688 正吻

13.870 正脊

6.810 飞檐

5.545 檐柱

0.880 下碱

±0.000 台明

-0.750 室外地坪

1818
7060
15688
1265
4665
880
750

8850 2215 3731 3909 3909 3718 1279
8850 18761

广胜下寺后殿梢间横剖面图

Cross Section of the Easternmost and Westernmost Bays of the Great Buddha Hall, Lower Guangsheng Monastery

0 1 3m

斗栱分布示意

斗栱类型	斗栱位置		1				2		3			4		
			东一	西一	东四	西四	平均值	东三	平均值	内东一	内西一	平均值	内东二	内西二	平均值
栌斗	①	490	500	492	494	494	495	495	432	442	437	556	558	557
	②	127	128	120	118	123	121	121	118	114	116	118	124	121
	③	290	292	272	285	285	275	275	255	290	273	290	295	293
	④	184	190	178	177	182	181	181	157	168	163	176	175	176
	⑤	430	435	432	433	433	433	433	404	440	422	460	464	462
	⑥	143	144	145	146	145	145	145	132	140	136	150	146	148
	⑦	148	146	148	147	147	149	149	134	135	135	151	148	150
华栱小斗	①	275	276	278	270	275	275	275	254	245	250	276	283	280
	②	60	59	59	60	60	55	55	55	39	47	55	54	55
	③	144	147	146	147	146	142	142	140	140	140	150	145	148
	④	90	89	87	90	89	85	85	85	77	81	82	89	86
	⑤	235	235	235	231	234	232	232	215	222	219	236	237	237
	⑥	148	142	149	143	146	141	141	140	150	145	147	147	148
泥道栱小斗	①	250	251	242	247	248	248	248	255	220	238	250	244	247
	②	57	60	58	57	58	58	58	55	50	53	55	72	64
	③	148	148	150	149	149	145	145	140	142	141	150	145	148
	④	88	88	90	89	89	89	89	85	75	80	82	89	86
	⑤	230	232	225	226	228	228	228	211	218	215	216	221	219
	⑥	147	148	144	145	146	146	146	145	142	144	147	148	148
慢栱小斗	①	246	256	252	250	251	250	250						
	②	56	54	54	51	54	55	55						
	③	140	150	148	145	146	145	145						
	④	88	92	90	87	89	85	85						
	⑤	230	232	226	224	228	228	228						
	⑥	136	140	137	139	138	138	138						
瓜子栱小斗	①	250	227	252	244	243	249	249						
	②	60	49	42	74	56	47	47						
	③	144	150	150	149	148	146	146						
	④	88	90	90	88	89	83	83						
	⑤	220	245	224	223	228	216	216						
	⑥	149	151	148	150	150	146	146						
慢栱（前）小斗	①	250	248	248	249	249	246	246						
	②	63	55	76	59	63	60	60						
	③	145	140	149	144	145	149	149						
	④	82	85	91	93	88	87	87						
	⑤	220	220	220	221	220	220	220						
	⑥	139	135	134	140	137	137	137						
令栱小斗	①	220	250	243	250	241	240	240						
	②	56	60	60	76	63	55	55						
	③	150	150	145	149	149	145	145						
	④	90	89	88	92	90	87	87						
	⑤	246	224	221	220	228	220	220						
	⑥	138	138	137	139	138	141	141						
昂小斗	①	276	272	277	274	275	275	275						
	②	58	74	58	57	62	65	65						
	③	145	144	148	149	147	149	149						
	④	88	93	95	92	92	88	88						
	⑤	230	236	224	230	230	231	231						
	⑥	149	149	150	148	149	148	148						
泥道栱、慢栱	①	224	218	216	220	220	229	229						
	②	406	383	381	397	392	382	382						
	③	296	316	312	302	307	312	312						
	④	311	307	302	312	308	302	302						
瓜子栱、慢栱（前）	①	240	194	207	245	222	235	235						
	②	381	394	365	379	380	388	388						
	③	297	297	293	292	295	283	283						
	④	306	307	323	290	307	314	314						
华栱	①								335	321	328	420	417	419
	②								273	284	279	297	311	304
泥道栱	①								352	327	340	429	450	440
	②								273	275	274	287	261	274
令栱	①	476	480	464	457	469	465	465						
	②	302	278	285	303	292	301	301						

广胜下寺后殿斗栱分布及数据表

Distribution Diagram and Statistics of Bracket Sets of the Great Buddha Hall, Lower Guangsheng Monastery

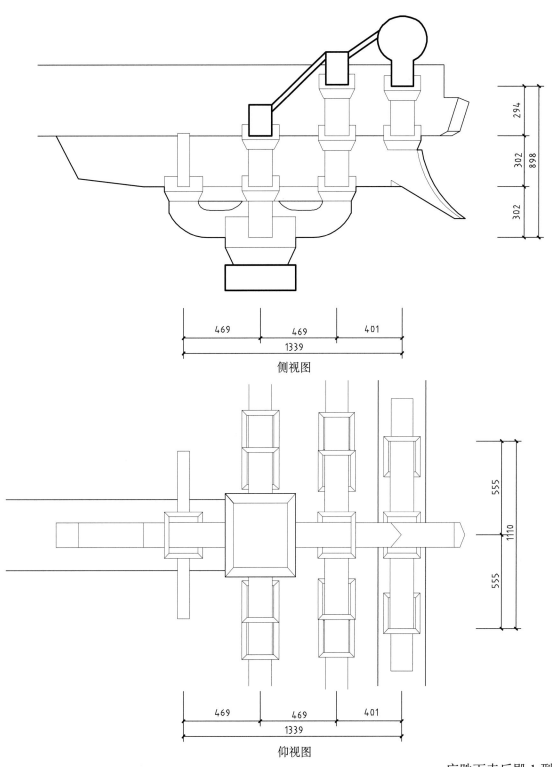

侧视图

仰视图

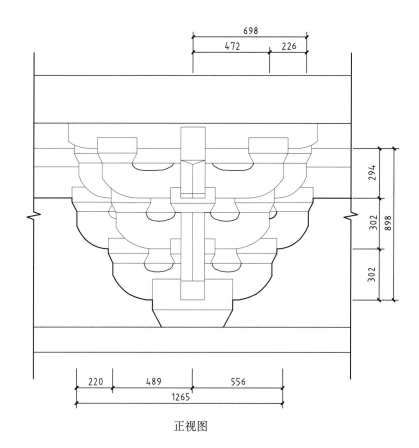

正视图

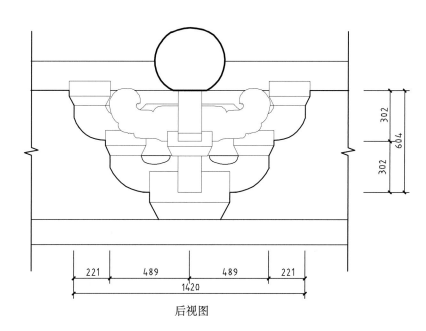

后视图

广胜下寺后殿 1 型斗栱大样图

Detailed Drawing of Bracket Sets Type 1 of the Great Buddha Hall, Lower Guangsheng Monastery

0 0.1   0.5m

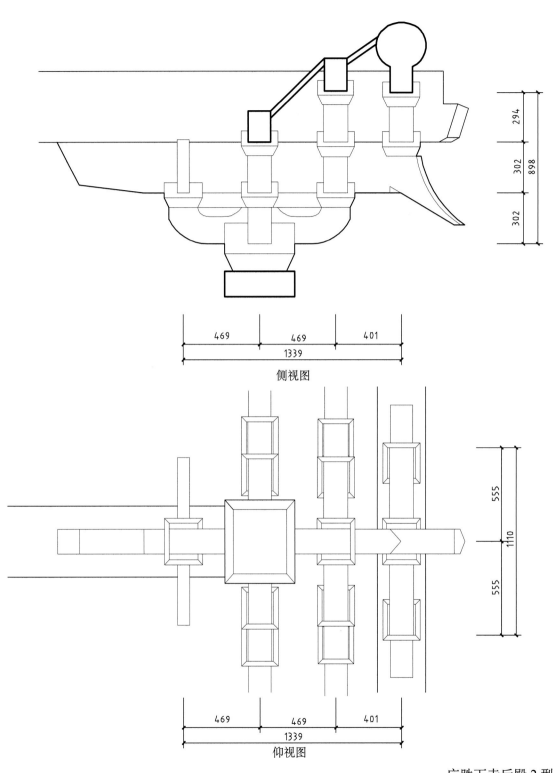

侧视图

仰视图

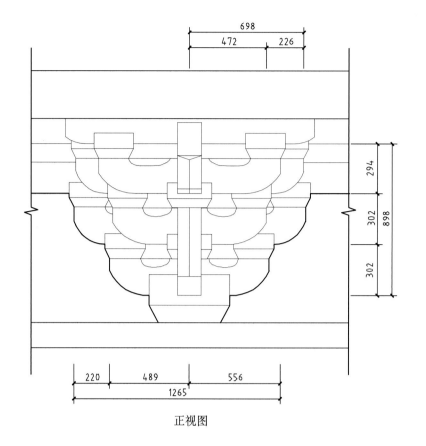

正视图

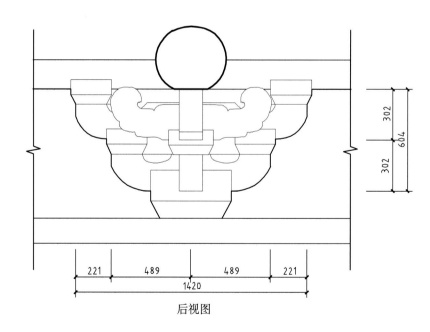

后视图

广胜下寺后殿 2 型斗栱大样图

Detailed Drawing of Bracket Sets Type 2 of the Great Buddha Hall, Lower Guangsheng Monastery

0   0.1   0.5m

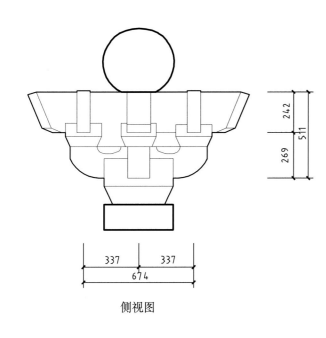

侧视图

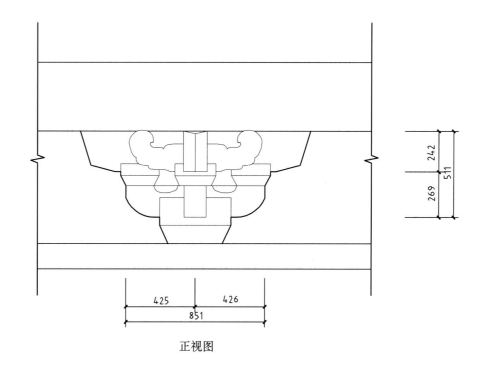

正视图

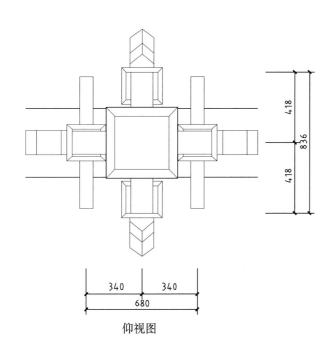

仰视图

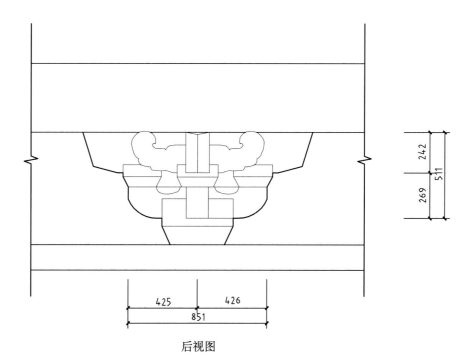

后视图

广胜下寺后殿 3 型斗栱大样图

Detailed Drawing of Bracket Sets Type 3 of the Great Buddha Hall, Lower Guangsheng Monastery

0　0.1　　　0.5m

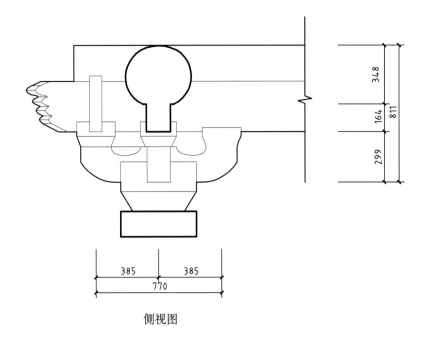

侧视图

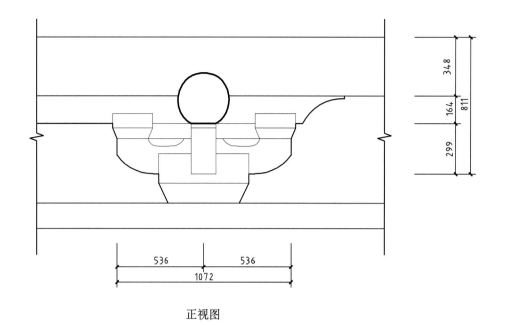

正视图

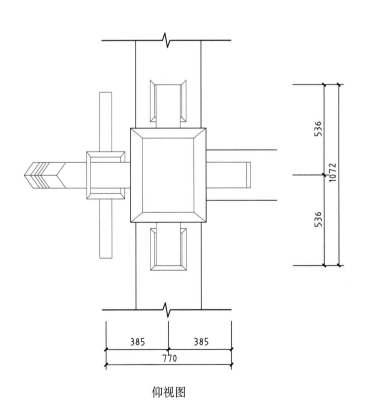

仰视图

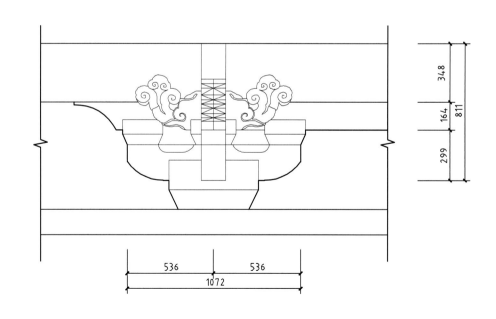

后视图

广胜下寺后殿 4 型斗栱大样图

Detailed Drawing of Bracket Sets Type 4 of the Great Buddha Hall, Lower Guangsheng Monastery

0   0.1   0.5m

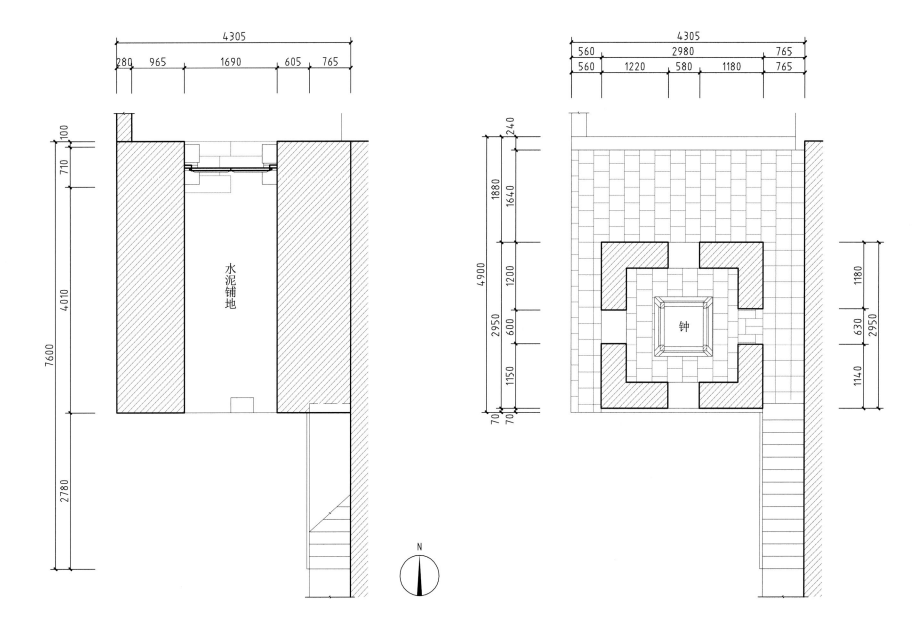

水泥铺地

钟

N

广胜下寺钟楼一层平面图
Ground Floor Plan of the Bell Tower, Lower Guangsheng Monastery

广胜下寺钟楼二层平面图
Second Floor Plan of the Bell Tower, Lower Guangsheng Monastery

0   0.5   1m

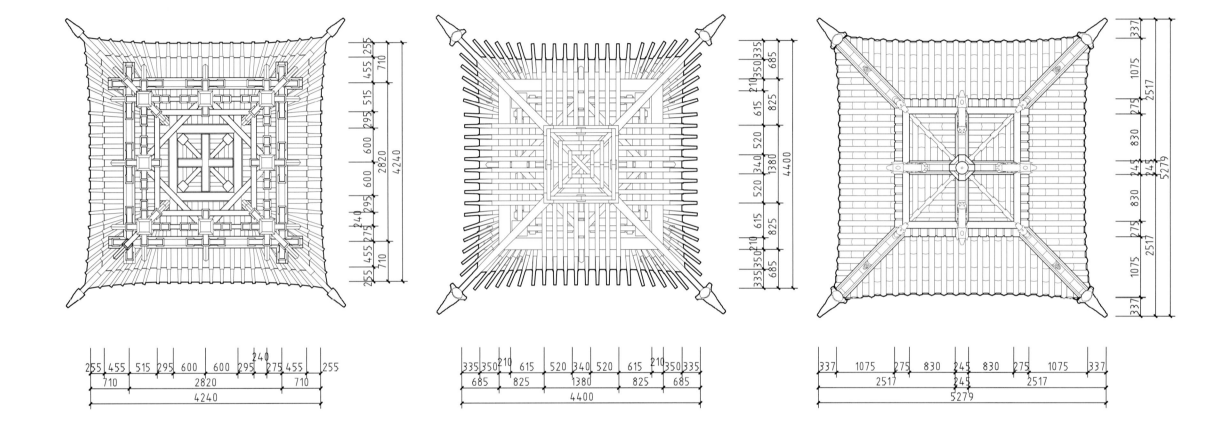

广胜下寺钟楼梁架仰视平面图
Bottom View of the Truss of the Bell Tower, Lower Guangsheng Monastery

广胜下寺钟楼梁架俯视平面图
Top View of the Truss of the Bell Tower, Lower Guangsheng Monastery

广胜下寺钟楼屋顶平面图
Top View of the Roof of the Bell Tower, Lower Guangsheng Monastery

0   0.5   1m

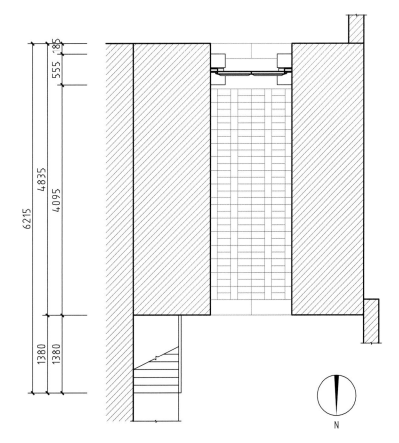

广胜下寺鼓楼一层平面图
Ground Floor Plan of the Drum Tower, Lower Guangsheng Monastery

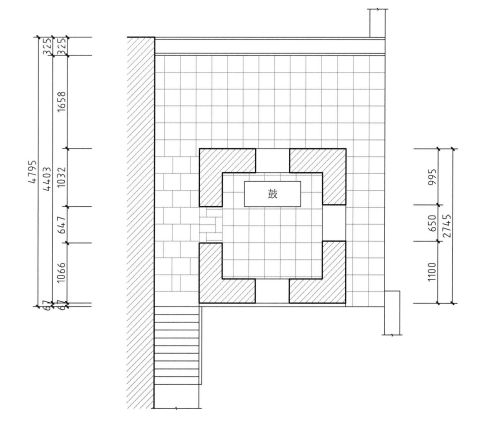

广胜下寺鼓楼二层平面图
Second Floor Plan of the Drum Tower, Lower Guangsheng Monastery

9 115 宝顶最高点

8 620 正吻最高点

8 254 正脊最高点

7 474 脊檩下皮

6 280 正身筒瓦上皮

6 124 正身飞椽上皮

4 695 茅顶

3 820 围栏最高点

2 940 二层地面

2 610

±0.000 台明

-0.060 地面

9115

60

2610 330 1755 520 909 156 1194 780 366 495 60

772 1736 964 356 585 941 585 356 964 1736 772

5355

90×90

90×90

120×80

120×80

Φ180

90×40

90×40

150×80

120×180

90×190

210×150

110×140

广胜下寺鼓楼剖面图

Section of the Drum Tower, Lower Guangsheng Monastery

0 0.5 1m

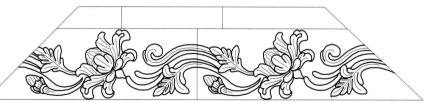

广胜下寺钟楼、鼓楼纹饰大样图
Detailed Drawing of the Ornamentation of Bell and Drum Towers, Lower Guangsheng Monastery

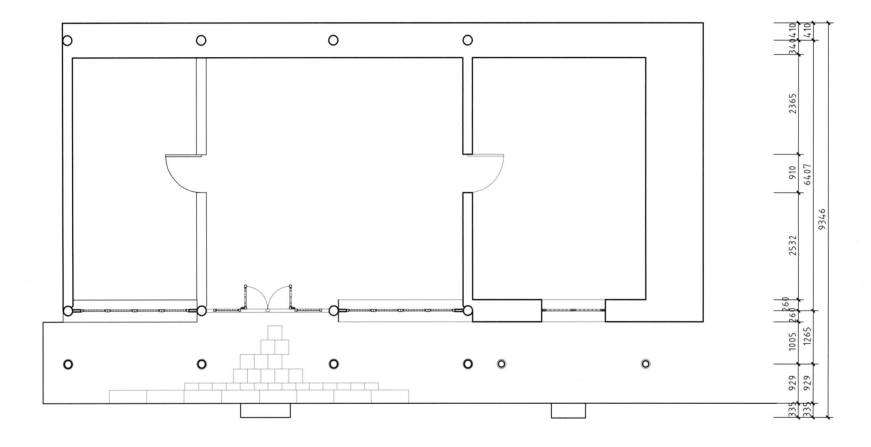

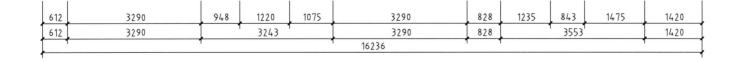

广胜下寺西配殿平面图
Floor Plan of the West Hall, Lower Guangsheng Monastery

0 1 2m

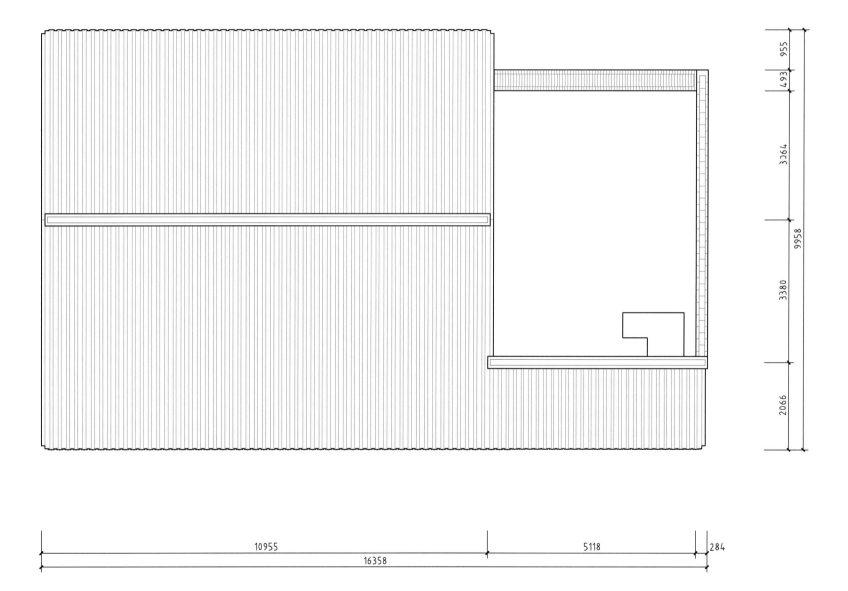

广胜下寺西配殿屋顶平面图
Top View of the Roof of the West Hall, Lower Guangsheng Monastery

0   1   2m

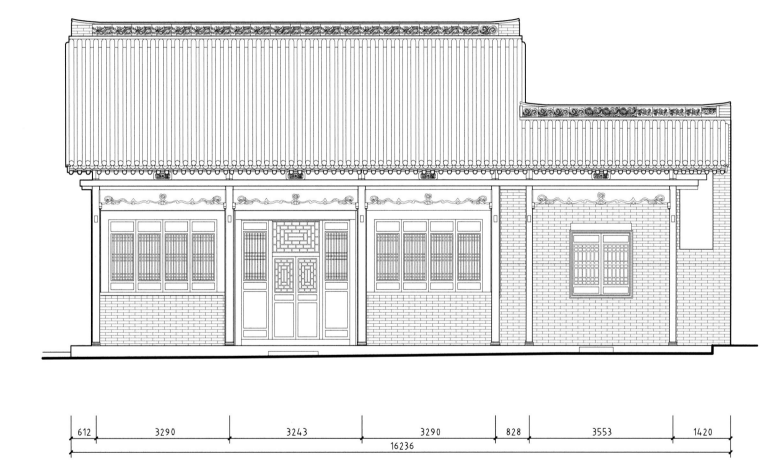

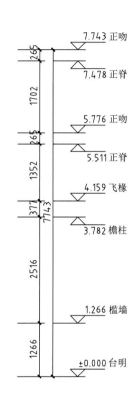

7.743 正吻

7.478 正脊

5.776 正吻

5.511 正脊

4.159 飞椽

3.782 檐柱

1.266 槛墙

±0.000 台明

265
1702
265
1352
377
7143
2516
1266

612　3290　3243　3290　828　3553　1420

16236

广胜下寺西配殿东立面图

Eastern Elevation of the West Hall, Lower Guangsheng Monastery

0　1　2m

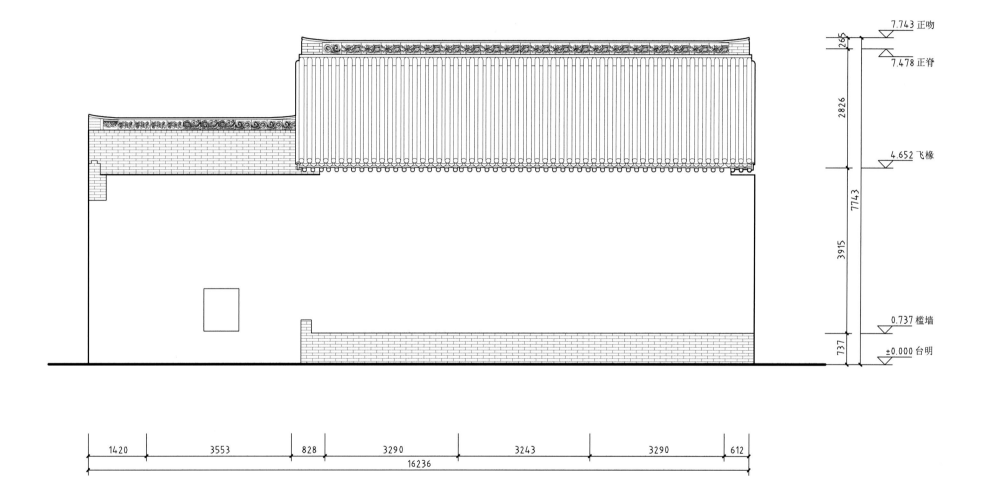

7.743 正吻
7.478 正脊
4.652 飞椽
0.737 槛墙
±0.000 台明

265
2826
7143
3915
737

1420　3553　828　3290　3243　3290　612
16236

广胜下寺西配殿西立面图
Western Elevation of the West Hall, Lower Guangsheng Monastery

0　1　2m

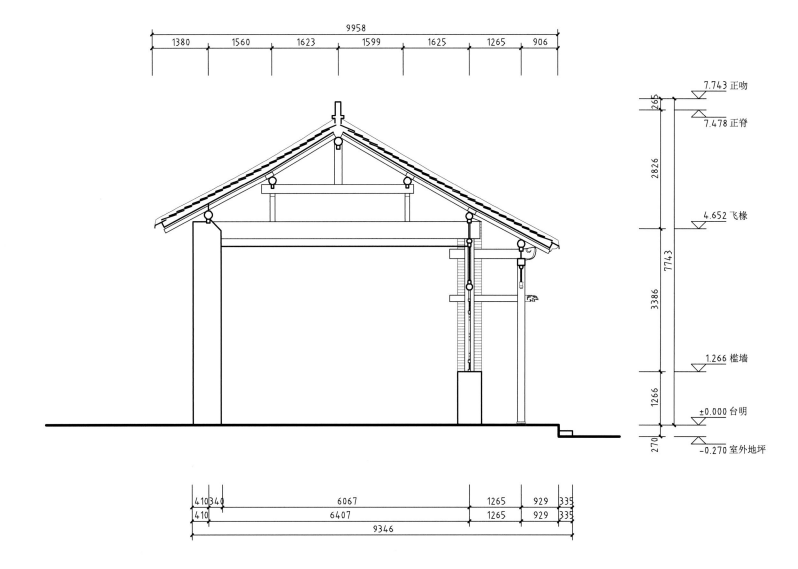

9958

| 1380 | 1560 | 1623 | 1599 | 1625 | 1265 | 906 |

265
7.743 正吻
7.478 正脊
2826
4.652 飞椽
7743
3386
1.266 槛墙
1266
±0.000 台明
270
-0.270 室外地坪

| 410 | 340 | 6067 | 1265 | 929 | 335 |
| 410 | | 6407 | 1265 | 929 | 335 |

9346

广胜下寺西配殿横剖面图
Cross Section of the West Hall, Lower Guangsheng Monastery

0 1 2m

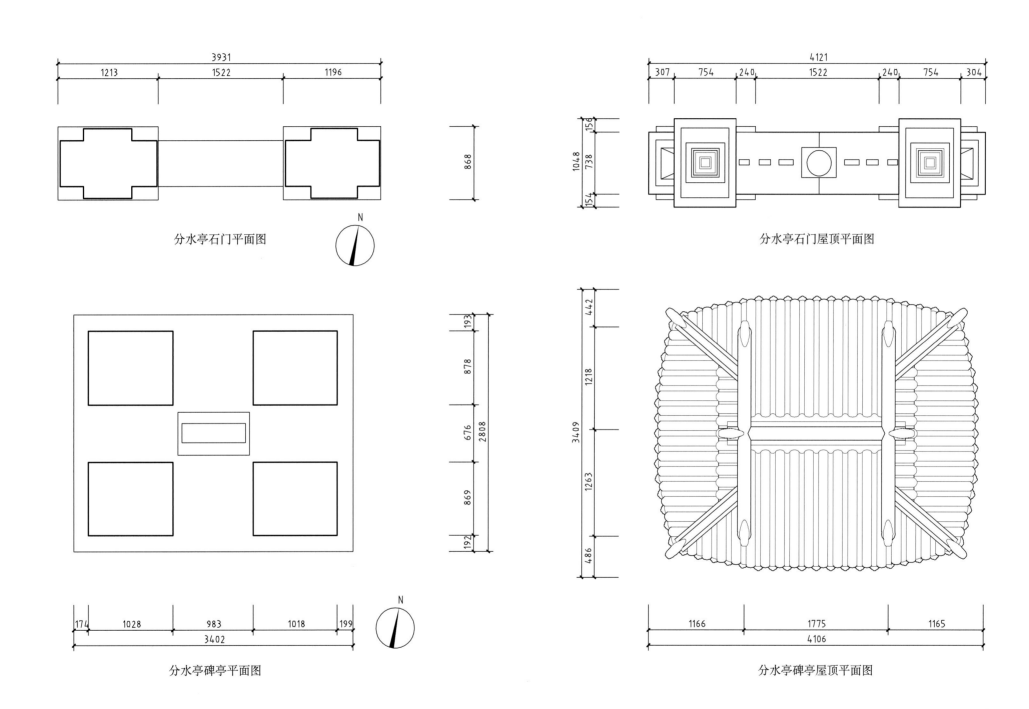

3931
1213　1522　1196
868

N

分水亭石门平面图

4121
307　754　240　1522　240　754　304
156　1048　738　154

分水亭石门屋顶平面图

193　878　676　2808　869　192

174　1028　983　1018　199
3402

N

分水亭碑亭平面图

442　1218　3409　1263　486

1166　1775　1165
4106

分水亭碑亭屋顶平面图

广胜下寺分水亭北侧石门及碑亭平面图、屋顶平面图
Floor Plan and Top View of Roofs of the Stone Gate and Stele Pavillion to the North of the Pavillion
of Water Allocation, Lower Guangsheng Monastery

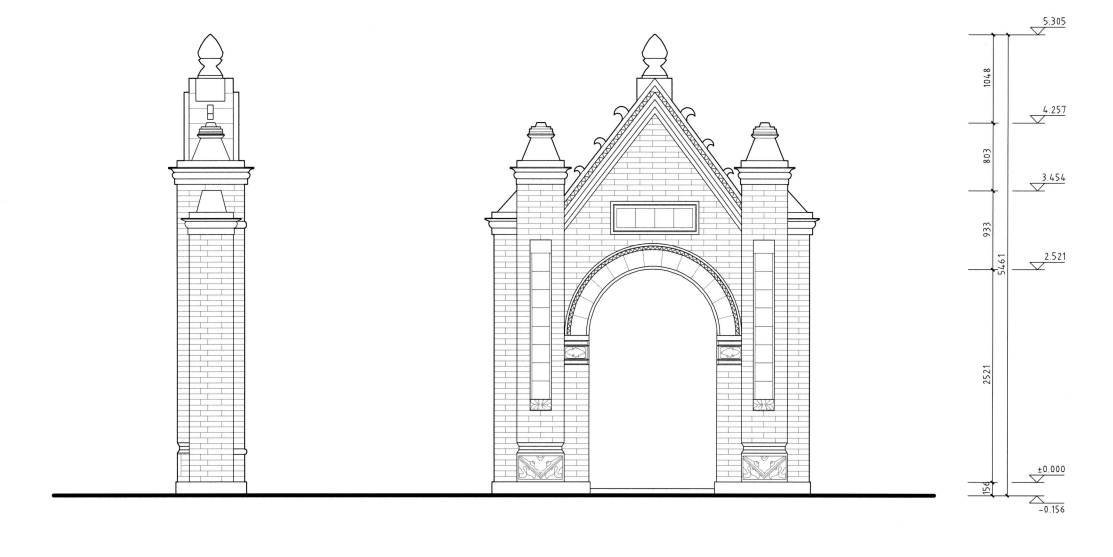

东立面图

南立面图

广胜下寺分水亭北侧石门立面图
Elevation of the Stone Gate to the North of the Pavillion of Water Allocation, Lower Guangsheng Monastery

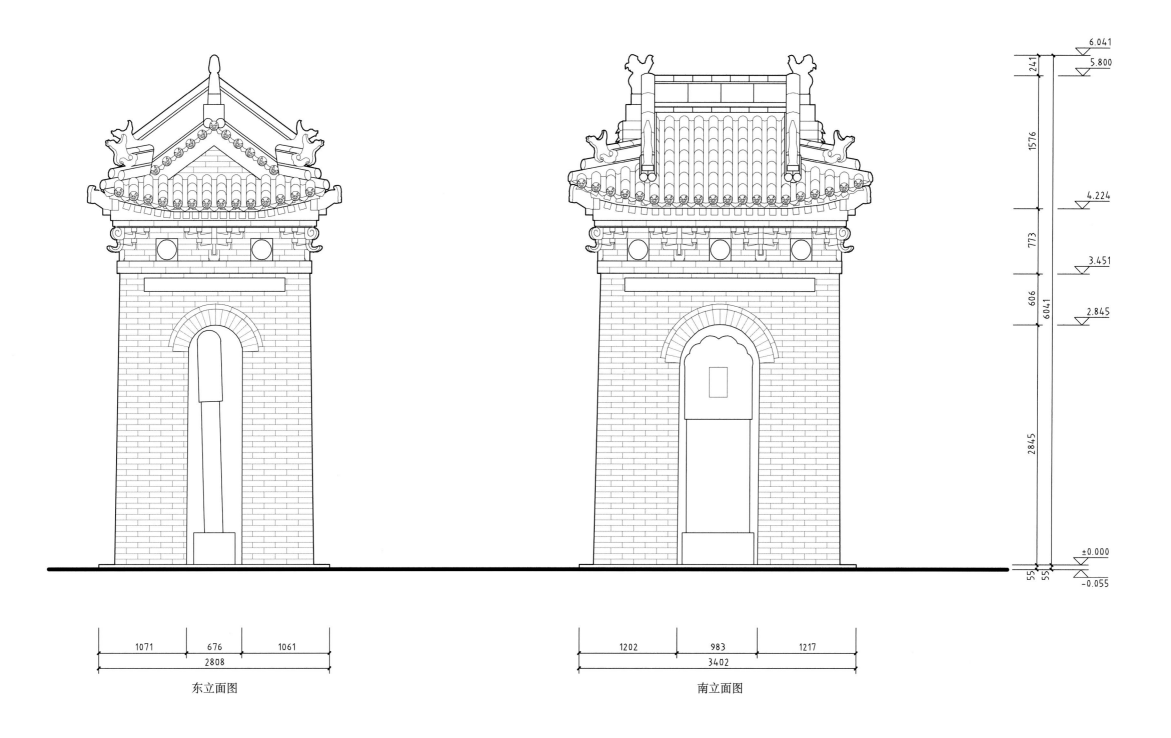

东立面图

南立面图

广胜下寺分水亭北侧碑亭立面图
Elevation of Stele Pavillion to the North of the Pavillion of Water Allocation, Lower Guangsheng Monastery

参与测绘及相关工作的人员名单

一、广胜上寺测绘人员
指导教师：廖慧农　杨　柳　王贵祥　刘　畅　黄文镐　刘梦雨　李　菁　王曦晨
测绘学生：陈　瑗　陈　茜　程思佳　王健南　孙晨炜　石圣松　杜顿康
　　　　　贾　珺　包媛迪
　　　　　时思芫　杨诗雨　张晨阳　吕诗旸　叶　晶　舒　畅　于　涛　曹嬿红
　　　　　谢殷睿　高菁辰　唐　丽　陆　达　齐轶昳　曹　越　张晗悠
　　　　　易斯坦　孙佳敏　李遇安　韦智宇　邓阳雪　高雅宁　吴雨凝
　　　　　韩靖北　杨天宇　陈　瑜　王子健　秦　岭　金恩惠　陆滢秀
　　　　　王静雯　谈家璐　吴朝赟　汪民权　黄若成　蔡泽宇

二、水神庙测绘人员
指导教师：李路珂　刘　畅　徐腾　赵波
测绘学生：徐　滢　许东磊　吕代越　齐大勇　金容辉　陈晓东　段俊毅　张成章
　　　　　王澜钦　郑　松　刘　涵　李嘉雨　唐　宁　蒋　哲　谢湘雅　秦正煜
　　　　　赵健程　谭婧玮　陈凌亚　朱钟晖

三、广胜下寺测绘人员
指导教师：刘　畅　刘梦雨　刘仁皓　贾　珺　徐　锋　徐　扬　李路珂
测绘学生：徐　滢　许东磊　吕代越　齐大勇　金容辉　郭　金　高祺　李新新
　　　　　甘旭东　杨　烁　冯　丹　刘冬元　金世雄　徐晨宇　殷玥　周辰
　　　　　刘凤逸　谢　骞　陈羚琪　王　冉　袁雪峰　薛昊天　马步青　卢清新
　　　　　张裕翔　黄　河　王澜钦　赵萨日娜

四、图纸整理及相关工作
图纸统筹：王　南　李　菁
图纸整理：匡天宇　陈德辰　马　傲　周敬砚　谢恬怡　吴　婧　闫芷宁
　　　　　荣　钰　雷宇芯　姜　明　李云开　杨　博　唐恒鲁　单梦林　买琳琳
　　　　　胡竞芙
英文统筹：[奥]荷雅丽
英文翻译：[奥]荷雅丽　Michael Norton　刘仁皓

Name List of Participants Involved in Surveying and Related Works

1. Surveying and Mapping of Upper Guangsheng Monastery

Supervising Instructor: LIAO Huinong, YANG Liu, WANG Guixiang, LIU Chang, HUANG Wenhao, LIU Mengyu, LI Jing, WANG Xichen, JIA Jun, BAO Aidi

Team Members: CHEN Ai, CHEN Xi, CHENG Sijia, WANG Jiannan, SUN Chenwei, SHI Shengsong, ZHANG Hanyou, DU Dikang, SHI Siyuan, YANG Shiyu, ZHANG Chenyang, LU Shiyang, YE Jing, SHU Chang, YU Tao, CAO Yuanhong, XIE Yinrui, GAO Jingchen, TANG Li, LU Da, QI Yiyi, CAO Yue, KANG Sidi, WU Yuning, YI Sitan, SUN Jiamin, LI Yu'an, WEI Zhiyu, DENG Yangxue, GAO Yaning, JIN Enhui, LU Yingxiu, HAN Jingbei, YANG Tianyu, CHEN Yu, WANG Zijian, QIN Ling, WANG Minquan, HUANG Ruocheng, CAI Zeyu, WANG Jingwen, TAN Jialu, WU Chaoyun

2. Surveying and Mapping of Water God's Temple

Supervising Instructor: LI Luke, LIU Chang, XU Teng, ZHAO Bo

Team Members: XU Ying, XU Donglei, LYU Daiyue, QI Dayong, JIN Ronghui, CHEN Xiaodong, DUAN Junyi, ZHANG Chengzhang, WANG Lanqin, ZHENG Song, LIU Han, LI Jiayu, TANG Ning, JIANG Zhe, XIE Xiangya, QIN Zhengyu, ZHAO Jiancheng, TAN Jingwei, CHEN Lingya, ZHU Zhonghui

3. Surveying and Mapping of Lower Guangsheng Monastery

Supervising Instructor: LIU Chang, LIU Mengyu, LIU Renhao, GU Jun, QING Feng, XU Yang, LI Luke

Team Members: XU Ying, XU Donglei, LU Daiyue, QI Dayong, JIN Ronghui, GUO Jin, GAO Qi, LI Xinxin, GAN Xudong, YANG Shuo, FENG Dan, LIU Dongyuan, JIN Shixiong, XU Chenyu, YIN Yue, ZHOU Chen, LIU Fengyi, XIE Qian, CHEN Lingqi, WANG Ran, YUAN Xuefeng, XUE Haotian, MA Buqing, LU Qingxin, ZHANG Yuxiang, HUANG He, WANG Lanqin, ZHAO Sarina

4. Editor of Drawings and Related Works

Drawings Arrangement: WANG Nan, LI Jing

Drawings Editor: KUANG Tianyu, CHEN Dechen, MA Ao, ZHOU Jingyan, XIE Tianyi, WU Jing, YAN Zhining, LI Hexin, RONG Yu, LEI Yuxin, JIANG Ming, LI Yunkai, YANG Bo, TANG Henglu, SHAN Menglin, MAI Linlin, HU Jingfu

Translator in Chief: Alexandra Harrer

Translation Members: Alexandra Harrer, Michael Norton, LIU Renhao

图书在版编目（CIP）数据

洪洞建筑群 = ARCHITECTURE COMPLEX OF HONGTONG：汉英对照 / 清华大学建筑学院编写；王贵祥等主编 . —北京：中国建筑工业出版社，2019.12
（中国古建筑测绘大系 . 宗教建筑）
ISBN 978-7-112-24559-8

Ⅰ . ①洪… Ⅱ . ①清… ②王… Ⅲ . ①宗教建筑—建筑艺术—洪洞县—图集 Ⅳ. ① TU–885

中国版本图书馆CIP数据核字（2019）第286223号

丛书策划 / 王莉慧
责任编辑 / 李 鸽 陈海娇
英文审稿 / ［奥］荷雅丽（Alexandra Harrer）
书籍设计 / 付金红
责任校对 / 姜小莲

中国古建筑测绘大系 · 宗教建筑
洪洞建筑群
清华大学建筑学院 编写
王贵祥 刘畅 贾珺 廖慧农 王南 主编
Traditional Chinese Architecture Surveying and Mapping Series: Religious Architecture
ARCHITECTURE COMPLEX OF HONGTONG
Compiled by School of Architecture, Tsinghua University
Edited by WANG Guixiang, LIU Chang, JIA Jun, LIAO Huinong, WANG Nan

*

中国建筑工业出版社出版、发行（北京海淀三里河路9号）
各地新华书店、建筑书店经销
北京方舟正佳图文设计有限公司制版
北京雅昌艺术印刷有限公司印刷

*

开本：787 毫米 × 1092 毫米 横 1/8 印张：29½ 字数：782 千字
2022 年 6 月第一版 2022 年 6 月第一次印刷
定价：**228.00** 元
ISBN 978-7-112-24559-8
（35133）